IEE RADAR, SONAR, NAVIGATION AND AVIONICS SERIES 6

Series Editors: Professor E. D. R. Shearman
P. Bradsell

SURFACE-
PENETRATING
RADAR

Other volumes in this series:

The cover shows the radar cross-sectional image of a steel reinforced concrete road. The area on the front between the blue horizontal lines represents a depth of 60 cm and a horizontal distance of 3m.

The key feature shown in the image is the ability of radar to propagate through a steel reinforcing mesh and still provide a clear image of the boundary between the sub-base of the concrete road and the base soil material (clay).

The image is reproduced by courtesy of ERA Technology and was taken with an SPRScan radar and 1 GHz antenna.

SURFACE-PENETRATING RADAR

D. J. Daniels
BSc MSc CEng MIEE MIEEE

The Institution of Electrical Engineers

Published by: The Institution of Electrical Engineers, London,
United Kingdom

© 1996: The Institution of Electrical Engineers

British Library Cataloguing in Publication Data

A CIP catalogue record for this book
is available from the British Library

ISBN 0 85296 862 0

Printed in England by Short Run Press Ltd., Exeter

Contents

Preface

This book describes the key elements of the subject of surface-penetrating radar, and in general terms the inter-relationship between those topics in electromagnetism, soil science, geophysics and signal processing which form a critical part of the design of a surface-penetrating-radar (SPR) system.

After an introduction in Chapter 1 to set the scene, the general system considerations are discussed in Chapter 2. Chapter 3 provides an introduction to the dielectric properties of earth materials. The characteristics of antennas suitable for use in SPR systems are described in Chapter 4 and are followed in Chapter 5 by a description of the various modulation techniques. Chapter 6 reviews the variety of signal-processing options currently available. Indications are given of the range of options available and ways in which various workers have approached their design and implementation for a given application are described. Most of the applications to which surface-penetrating radar has been put in recent years are reviewed in Chapter 7, and Chapter 8 discusses methods of field survey and operation and describes a representative selection of the currently commercially available equipment.

The aim of this book is to provide an introduction for the newcomer to the field, as well as a useful source of further reading, information and references for the current practitioner. The objective in writing this book is to bring together in one volume all the core information on a technique which spans a wide range of disciplines. While much of this is available in a range of different publications, it is dispersed and therefore less accessible by virtue of the disparate nature of many of the sources.

By necessity, this book provides a snapshot of the field of surface-penetrating or ground-penetrating radar, and it is to be expected that further developments in hardware and signal-processing techniques will incrementally improve the performance and extent of applications. Most companies manufacturing radar equipment have been helpful in supplying information, and Chapter 8 thus represents the state-of-the-art at the time of writing.

If this book helps the newcomer to assess the potential of the technique correctly and apply it effectively, its purpose will be well served.

An example may illustrate the reason for the previous comment. Several years ago a suggestion was made that a particular ground-probing radar and its operator could detect targets the size of golf balls at a depth of 8 m. Clearly, the wavelengths capable of propagating to 8 m would be so much larger than a golfball-sized target that the radar cross-sectional area of the latter would fade into insignificance.

The persuasiveness of the claimant and the lack of understanding of basic physics on the part of the potential users enabled this kind of claim to be seriously proposed.

Therefore, a secondary objective in preparing this volume has been to provide a source of information which will allow potential users to assess the merits of claims, sound or otherwise.

Surface-penetrating radar is, like all other engineering techniques, firmly based on physical principles which must be understood if the technique is to be properly applied. In reality, a metre of wet clay is still largely opaque, even to the latest radar hardware, however well provisioned with arrays of microprocessors, artificial intelligence and neural networks.

It should also be noted that it is not the intention to address the expert or advanced practitioner in the discipline, as to some extent they have contributed either by reference or by inclusion of material. However, if this book provides a source of reference it may be found useful by those with significant expertise.

The treatment of the subject is generally at that of first-year undergraduate level, although some chapters may require a deeper knowledge of antenna and EM-wave theory. The aim has been to provide a treatment which is readily accessible, rather than a volume for the advanced specialist, who I feel is well capable of assessing the latest research papers. As such, I have chosen to provide information which has at least been proven by application.

For the survey operator of a radar, the moment of truth arrives when, while on-site, the client looks at the radar image and requests an on-line interpretation of what the images represent and more importantly where and how deep to dig. Nothing less than a sound understanding will enable a correct decision to be made and proper advice to be given.

Acknowledgments

I am extremely grateful for the help that I received from many fellow workers in the field of ground-probing radar. In particular, I would like to thank those who provided individual contributions to Chapter 7 on Applications: Dr Manfred Bartha, Mr H. F. Scott, Dr J. Bungey, Dr R. de Vekey, Dr P. Hanninen, Dr van Overmeeren, Dr J. K. van Deen, Dr S. Tillard, Dr J. Fidler, Mr D. L. Wilkinson, Dr J. Cariou, Dr G. Schaber, Prof. T. Haglars, Ms F. Nicollin; and to Chapter 9 on Equipment of the following companies: Greg Mills of GSSI (US), Sensors and Software (Canada), Redifon (UK), ERA Technology (UK) NTT (Japan) and MALA (Sweden).

I would also like to express my appreciation of discussions over the years with many other engineers and scientists working in the field. Where possible, I have cited their publications and if referencing has been omitted I am pleased to acknowledge the contribution everyone has made to developing a specialist area of radar technology.

I would also like to express my thanks and appreciation to the Directors of ERA Technology Ltd, UK, both for their support of research and development of the technology and for their assistance with the preparation of the volume. I would also like to thank various colleagues at ERA Technology who provided many useful comments on the manuscript. I am particularly grateful to Pam Wheeler who enthusiastically and patiently typed the manuscript and to Shirley Vousden who prepared all of the figures.

Finally, I would also like to thank my wife, Jenny, for her support and encouragement during the time it took to write this book.

List of symbols used

The following notation has been employed:

English

c	= velocity of electromagnetic waves in free space $(3.10^8 \text{m.s.}^{-1})$	m/s
d	= depth	m
E	= instantaneous electric-field component of EM wave	V/m
E_0	= original electric field $(z=0, t=0)$ of propagating EM wave	V/m
$E(z,t)$	= electric field at z, t of propagating EM wave	V/m
f	= frequency	Hz
H	= instanteous magnetic-field component of EM wave	A/m
j	= $\sqrt{-1} = i$	
k	= phase constant (can also be referred to as wave number)	
x	= cartesian co-ordinate-axis value	
y	= cartesian co-ordinate-axis value	
z	= cartesian co-ordinate-axis value	
R or r	= range of target from transmitter	m
P_r	= received power	W
P_t	= transmitted power	W
G_t	= gain of transmit antenna	(dimensionless)
G_r	= gain of receive antenna	(dimensionless)
A_r	= aperture of receive antenna	m^2
B	= bandwidth	Hz
q	= charge	C
$f(t)$	= function of time	
$f(\omega)$	= function of frequency	
F	= Fourier transform	
F^{-1}	= inverse Fourier transform	
N	= number of averages	
n	= sample number	

Greek

α	= attenuation constant	m^{-1}
α'	= attenuation constant in dB/m	dB/m

β	= phase constant	m^{-1}
$\tan \delta$	= material loss tangent = ϵ''/ϵ'	
ϵ	= absolute permittivity of medium	F/m
ϵ_0	= absolute permittivity of free space	F/m
ϵ_r	= relative permittivity of medium	F/m
ϵ'	= real part of permittivity of medium	F/m
ϵ''	= imaginary part of permittivity of medium	F/m
ϵ_s	= low-frequency (static) permittivity	F/m
ϵ_∞	= high-frequency (infinite) permittivity	F/m
ϵ_m	= permittivity of medium	F/m
ϵ_w	= permittivity of water	F/m
η	= intrinsic impedance of medium	Ω
λ	= wavelength of EM waves in medium	m
τ	= relaxation time of water	s
ω	= $2\pi f$	rad/s
μ	= absolute magnetic permeability of medium	H/m
μ_0	= absolute magnetic permeability of free space	H/m
μ_r	= relative magnetic permeability of medium	H/m
π	= pi	
σ	= absolute conductivity of medium	S/m
σ_{DC}	= DC or ohmic conductivity of medium	S/m
σ'	= real part of conductivity of medium	S/m
σ''	= imaginary part of conductivity of medium	S/m
ν	= velocity of propagation of EM wave in air	m/s
ν_r	= velocity of propagation of EM waves in material	m/s
δ	= loss factor in material	

Chapter 1
Introduction

1.1 Overview

The possibility of detecting buried objects remotely has fascinated mankind over centuries. A single technique which could render the ground and its contents clearly visible is potentially so attractive that considerable scientific and engineering effort has gone into devising suitable methods of exploration.

As yet, no single method has been found to provide a complete answer, but seismic, electrical-resistivity, induced-polarisation, gravity-surveying, magnetic-surveying, nucleonic, radiometric, thermographic and electromagnetic methods have all proved useful. Ground-probing or surface-penetrating radar has been found to be a specially attractive option. The subject has a special appeal for practising engineers and scientists in that it embraces a range of specialisations such as electromagnetic-wave propagation in lossy media, ultra-wideband-antenna technology and radar-systems design, discriminant-waveform signal processing and image processing. Most surface-penetrating radars are a particular realisation of ultra-wideband impulse-radar technology. Skolnik (1990) considers that 'the technology of impulse radar creates an exciting challenge to the innovative engineer' and ground-probing radar to have been a successful commercial venture although on a smaller scale than conventional radar applications.

The terms ground-probing radar, subsurface radar or surface-penetrating radar refer to a range of electromagnetic techniques designed primarily for the location of objects or interfaces buried beneath the earth's surface or located within a visually opaque structure. The term surface-penetrating radar is adopted by the author as it describes most accurately the application of the method to the majority of situations including buildings, bridges etc, as well as probing through the ground. The technology of surface-penetrating radar is largely applications-oriented and the overall design philosophy, as well as the hardware, is usually dependent on the target type and the material of the target and its surroundings. The range of applications for surface-penetrating-radar methods is wide and the sophistication of signal-recovery techniques, hardware designs and operating practices is increasing as the technology matures. More recent developments include airborne and satellite surveying as well as high-speed survey from vehicle-mounted radars.

This book is intended as an introduction on the subject and is aimed at the growing number of potential users who wish to gain an introductory understanding of the method at a level appropriate to the first year of undergraduate studies. It is assumed that, whatever the reader's background, be it in geophysics, civil engineering or archaeology, he or she has a basic understanding of physics and geophysics and understands that radar is an active measurement technique which allows the ranging and detection of targets. Although each Chapter has been written in the form of a self-contained section relevant to its particular topic, the overall aim of the author has been to create a sufficiently wide-ranging treatment that will enable interested readers to investigate areas of particular interest in greater depth. To assist the reader, a list of suitable references is provided at the end of each Chapter.

Following early laboratory developments in the late 1960s and 1970s in both the US and UK, commercial equipment has become more freely available and the GSSI impulse radar has become the most widely used commercial system. More recently, alternative equipment using various modulation techniques has become available and the market is expanding. However, the impulse radar has been the most successful design to date and probably accounts for 95% of the units in operation in the field.

The general structure of this volume is based on an earlier paper which was published in a special edition of the Proceedings of the IEE Part F (Daniels *et al.*, 1988) which served as an introduction to surface-penetrating-radar techniques. This still serves well as a primer, and the introduction is still relevant and is quoted below.

1.2 History

The first use of electromagnetic signals to determine the presence of remote terrestrial metal objects is generally attributed to Hülsmeyer in 1904, but the first description of their use for location of buried objects appeared six years later in a German patent by Leimbach and Löwy. Their technique consisted of burying dipole antennas in an array of vertical boreholes and comparing the magnitude of signals received when successive pairs were used to transmit and receive. In this way, a crude image could be formed of any region within the array which, through its higher conductivity than the surrounding medium, preferentially absorbed the radiation. These authors described an alternative technique which used separate, surface-mounted, antennas to detect the reflection from a subsurface interface due to ground water or to an ore deposit. An extension of the technique led to an indication of the depth of a buried interface, through an examination of the interference between the reflected wave and that which leaked directly between the antennas over the ground surface. The main features of this work, namely continuous-wave (CW) operation, use of shielding or diffraction effects due to underground features, and the reliance on conductivity variations to produce scattering, were present in a number of other patent disclosures, including some intended for totally submerged applications in mines. The work of Hülsenbeck (1926) appears to be the first use of pulsed techniques to determine the structure of buried features. He noted that any dielectric variation, not necessarily involving conductivity, would also produce reflections and that the technique, through the easier realisation of directional sources, had advantages over seismic methods.

Pulsed techniques were developed from the 1930s onwards as a means of probing to considerable depths in ice (Steenson, 1951; Evans, 1963), fresh water, salt deposits, desert sand and rock formations (Morey, 1974; Cook, 1974; Unterberger, 1978; Kadaba, 1976; Cook, 1975; Roe and Ellerbruch, 1979; Nilsson, 1978; McCann *et al.*, 1988). Probing of rock and coal was also investigated by Cook (Cook, 1974; Cook, 1975), although the higher attenuation in the latter material meant that depths greater than a few metres were impractical. A range of civil-engineering applications has been reviewed by Morey (1974). A more extended account of the history of surface-penetrating radar and its growth up to the mid-1970s is given by Nilsson (1978).

Renewed interest in the subject was generated in the early 1970s when lunar investigations and landings were in progress. For these applications, one of the advantages of subsurface radar over seismic techniques was exploited, namely the ability to use remote, noncontacting transducers of the radiated energy, rather than the ground-contacting types needed for seismic investigations. Remote transducers are possible because the dielectric impedance ratio between free space and soil materials, typically from 2 to 4, is very much less than the corresponding ratio for acoustic impedances, by a factor which is typically of the order of 100.

From the 1970s until the present day, the range of applications has been expanding steadily, and now includes building and structural nondestructive testing, archaeology, road and tunnel quality assessment, location of voids and containers, tunnels and mineshafts, pipe and cable detection, as well as remote sensing by satellite. Purpose-built equipment for each of these applications is being developed, and the user now has a better choice of equipment and techniques.

1.3 Applications

Recent progress has been one of continuing technical advance largely applications-driven, but as the requirements have become more demanding, so the equipment, techniques and data-processing methods have been developed and refined.

The main operational advantages of the technique can be described as follows. The antennas of a surface-penetrating radar do not need to be in contact with the surface of the earth, thereby allowing rapid surveying. Indeed, surface-penetrating radars developed by Stanford Research Institute have been flown at heights of 400 m in synthetic-aperture radar (SAR) mode and have imaged buried metallic mines.

Antennas may be designed to have adequate properties of bandwidth and beam shape, although optimum performance, especially where a small antenna-to-ground surface spacing is involved, will usually be obtained only by taking into account details of the geometry and the nature of the ground.

Signal sources are available which can produce subnanosecond impulses or, alternatively, which can be programmed to produce a wide range of modulation types.

In general, any dielectric discontinuity is detected. Targets can be classified according to their geometry: planar interfaces; long, thin objects; localised spherical or cuboidal objects. The radar system can be designed to detect a given target type preferentially and is potentially capable of producing an image of the target in three dimensions, although little work has been done in this aspect of image presentation.

The signal attenuation at the desired operating frequency is the main factor to be considered when assessing the usefulness of radar probing in a given material. As a rule, material which has a high low-frequency conductivity will have a large signal attenuation. Thus gravel, sand, dry rock and fresh water are relatively easy to probe using radar methods, while salt water, clay soils and conductive ores or minerals are less so, but a reduction in the transmitted frequency means that even these materials can be adequately investigated, though at the expense of a reduced resolution between targets. Surface-penetrating radar will work successfully in fresh water so that water content is not a complete guide to achievable penetration range.

Surface-penetrating radar relies for its operational effectiveness on meeting successfully the following requirements: (*a*) efficient coupling of electromagnetic radiation into the ground; (*b*) adequate penetration of the radiation through the ground having regard to target depth; (*c*) obtaining from buried objects or other dielectric discontinuities a sufficiently large scattered signal for detection at or above the ground surface; and (*d*) an adequate bandwidth in the detected signal having regard to the desired resolution and noise levels.

The essence of the technique is no different from that of conventional, free-space radar, but of the factors which affect the design and operation of any radar system the four requirements indicated earlier take on an additional significance in subsurface-radar work. Specifically, propagation loss, clutter characteristics and target characteristics are distinctly different.

The radar technique is usually employed to detect backscattered radiation from a target. Forward scattering can also yield target information, although for subsurface work at least one antenna would need to be buried, and an imaging transform would need to be applied to the measured data.

The designer of radar for subsurface applications has two problems not necessarily encountered by the designer of a conventional radar: designing to a limited budget and overcoming difficult signal-recovery problems. As the cost of processing falls, the designer can consider signal-processing strategies previously thought uneconomic and this will be likely to have a significant effect on system design and hence commercial viability.

For example, in early radars for subsurface applications, it was considered necessary to employ wideband antennas with linear phase response, because of the resulting difficulties in deconvolving the antenna response. However, it is now possible to correct for nonlinear phase characteristics, if desired, at a reasonable cost by using appropriate signal-processing techniques implemented in software.

Surface-penetrating radar is vulnerable to extremely high levels of clutter at short ranges and this rather than signal/noise recovery is its major technical handicap. The system to be specified should take this into account. Some basic guidelines can be suggested for the user of radar for ground-probing applications.

It is important to define clearly the target parameters. There is a considerable difference between the target responses from a buried pipe, a buried mine, a void and a planar interface. This has a major impact on antenna design, polarisation state and signal-processing strategy, and should be exploited.

The resolution and depth requirement needed should be clearly identified. This, in turn, sets the frequency and bandwidth of operation, which then influence the choice of modulation technique and hence the hardware design. The costs of overspecification can be considerable, and the physics of propagation should be kept in mind.

The transmission-loss characteristics will affect the selection of a system. It is unlikely that synthetic-aperture or holographic schemes will work well in high-loss materials.

The display requirements can have a major impact on equipment costs. This has a fundamental bearing on the type of signal processing required. Image presentation obviously needs different signal processing from that required for target identification and classification. Signal processing must take account of the needs of the user, and as much as possible of the interpretation process should be done automatically, if surface-penetrating-radar techniques are to gain widespread acceptance for routine use in, say, pipe and cable location.

The operational requirements are such that physical decoupling of the antenna from the surface should only be carried out where strictly necessary. This is not simply for reasons of power transfer but also for reduction of clutter and efficiency of transfer.

Some of the ancillary requirements of an operational subsurface radar system need more consideration. There is a need for an accurate, small-scale, low-cost position-referencing system for use with radar for subsurface survey techniques. For utilities it will be most important that data can be related to a true geographic reference, particularly when filed on digital mapping systems and used to define areas of safe working. It will be necessary to provide some means of scanning the antenna. Obviously a basic approach is the handheld device, but this places severe limitations on the signal-processing strategies. Alternatives are robotic arms and miniature tracked vehicles; the former may limit the area of search but may be cheaper for surveying road and pavement areas, while the latter may be the most flexible, but will require accurate position referencing.

If the radar is to provide its operator with an image of targets under the ground surface, online processing will be needed. With the projected developments in advanced microprocessors, it is likely that significant amounts of online processing will soon become economically feasible.

1.4 Development

The key future development area will be signal-processing and image-recognition models, and this requires development of core strategies generally based on deconvolution techniques.

The future of surface-penetrating radar is considered to be based on short-range geophysical exploration and nondestructive investigation. For short-range geophysical exploration, subsurface radar has already achieved some significant results. It is, however, in the area of nondestructive investigation of structures such as tunnels, roads, buildings and other examples of physical infrastructure of modern civilisation that surface-penetrating radar has an increasingly important role to play.

Potential customers could be energy and communications utilities, mineral-resource-exploration organisations, civil-engineering organisations, nondestruc-tive-testing companies, military and security organisations, architects, archae-ologists and scientific research establishments. Many of these organisations only wish to purchase surface-penetrating radars provided that the price is within reasonable limits, and may prefer to hire the services of a specialist surveying organisation or, alternatively, hire equipment. It is likely that the commercial definition of a reasonable price for either commercial equipment, hire or service, will be different from the military definition of a reasonable price for a radar.

In addition, the experience of many of those commercial organisations in relation to electronic equipment is of mainstream suppliers of conventional equipment. Suppliers of production quantities achieve economies of scale which are unlikely to apply for a radar for subsurface applications.

The designer of surface-penetrating radars must therefore take into account the types of markets which exist at present. However, the success of the current commercial radars is encouraging and suggests that evolutionary design processes could widen the market.

One possible design strategy for surface-penetrating radar could be seen as the development of a modular system. With this approach, frequency range, and hence antenna type, can be modified simply and signal processing can be selected by choice of target and display format. This would allow economy in development and of production and result in a more commercially attractive product.

The potential cannot be overlooked of increasingly powerful microprocessors available at low cost, capable of carrying out sophisticated signal processing and then displaying the results so that interpretation by an expert is less important. This potential should provide the technology necessary for radar for subsurface applications to gain wide acceptance as a valuable investigative tool capable of being used by the nonspecialist. However, the pattern-recognition capability of the human brain is still unequalled and may remain so for many decades.

Surface-penetrating radar is one of a very few methods available which allows the inspection of objects or geological features which lie beneath an optically opaque surface. Much funding to date has come from industries in the civil sector with a direct interest in the information that can be derived from subsurface radar exploration: the utilities (gas, electricity, water, telecommunications), oil and gas exploration companies, geophysical-survey groups. The total expenditure is still small when compared with the investment, largely from military sources, in free-space-radar developments. However, it is generally considered that surface-penetrating radar has been a successful commercial venture even if its market-sector value is not as large as some of the radar applications.

With improvements in the performance of subsurface-radar systems will come wider commercial acceptance. The challenge for the designer is to speed up the rate of development. Subsurface-radar technology will become more firmly established as its benefits are perceived and realised by users distinctly different from those of conventional free-space-radar technology.

Although surface-penetrating radar has achieved some spectacular successes, it would be unrealistic to leave the impression that surface-penetrating radar is the complete solution to the users perceived problem (whatever that may be). A surface penetrating radar will detect, within the limits of the physics of propagation, all changes in electrical impedance in the material under investigation. Some of these changes will be associated with wanted targets; others may not be. The radar has, in general, no way of discriminating and much of the skill of the successful user at present comes from forming a conclusion from both the radar image and site intelligence. The more successful operators routinely exercise this discipline and procedure.

The potential user should therefore understand both the capabilities and limitation of the method. This volume will have achieved that objective if the user employs radar in the right place at the right time in parallel with other geophysical exploration methods.

1.5 References

COOK, J. C. (1974): Status of ground-probing radar and some recent experience. Proceedings of a conference on *Subsurface Exploration for Underground Excavation and Heavy Construction*, American Society of Civil Engineers, 175–194
COOK, J. C. (1975): Radar transparencies of mine and tunnel rocks. *Geophys.*, 40, 865–885
DANIELS, D. J., GUNTON, D. J., and SCOTT, H. F. (1988): Introduction to subsurface radar. *IEE Proc. F*, **135**, (4), 278–321
EVANS, S. (1963): Radio techniques for the measurement of ice thickness. *Polar Rec.*, **11**, 406-410
HÜLSENBECK, & CO (1926): German patent 489 434
KADABA, P. K. (1976): Penetration of 1 GHz to 1.5 GHz electromagnetic waves into

the earth surface for remote sensing applications. Proceedings of IEEE SE Region 3 conference, 48–50

McCANN, D. M., JACKSON, D. J., and FENNING, P. J. (1988): Comparison of the seismic and ground probing methods in geological surveying. *IEE Proc. F*, **135**, 380–390

MOREY, R. M. (1974): Continuous sub-surface profiling by impulse radar. Proceedings of conference on *Subsurface Exploration for Underground Excavation and Heavy Construction*, American Society of Civil Engineers, 213–232

NILSSON, B. (1978): Two topics in electromagnetic radiation field prospecting. Doctoral thesis, University of Lulea, Sweden

ROE, K. C., and ELLERBRUCH, D. A. (1979): Development of a microwave system to measure coal layer thickness up to 25 cm. US National Bureau of Standards, Boulder, CO, USA, report SR–723–8–79

SKOLNIK, M. I. (1990): An introduction to impulse radar. Naval Research Laboratory report 6755

STEENSON, B. O. (1951): Radar methods for the exploration of glaciers. PhD thesis, California Institute of Technology, Pasadena, CA, USA

UNTERBERGER, R. R. (1978): Radar and sonar probing of salt. Fifth international symposium on *Salt*, Hamburg, Northern Ohio Geological Society, USA, 423–437

Chapter 2
System design

2.1 Introduction

Surface-penetrating radar has an enormously wide range of applications, ranging from planetary exploration to the detection of buried mines. The selection of a range of frequency operation, a particular modulation scheme and the type of antenna and its polarisation depends on a number of factors including the size and shape of the target, the transmission properties of the intervening medium and the operational requirements defined by the economics of the survey operation, as well as the characteristics of the surface. The specification of a particular type of system can be prepared by examining the various factors which influence detectivity and resolution.

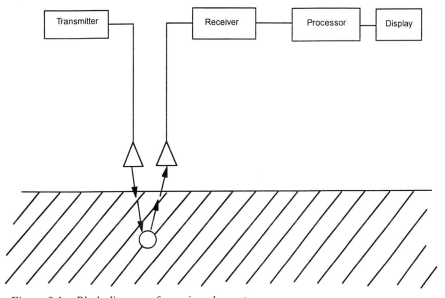

Figure 2.1 Block diagram of generic radar system

To operate successfully, surface-penetrating radar must achieve

(*a*)   an adequate signal-to-clutter ratio
(*b*)   an adequate signal-to-noise ratio
(*c*)   an adequate spatial resolution of the target
(*d*)   an adequate depth resolution of the target.

Most surface-penetrating-radar systems detect the backscattered signal from the target, although forward-transmission methods are used in borehole tomographic radar imaging.

This Chapter considers the principal factors affecting the design of a surface-penetrating radar in order to illustrate those factors which need to be considered. The aim is to illustrate the technical options available to the operator or designer.

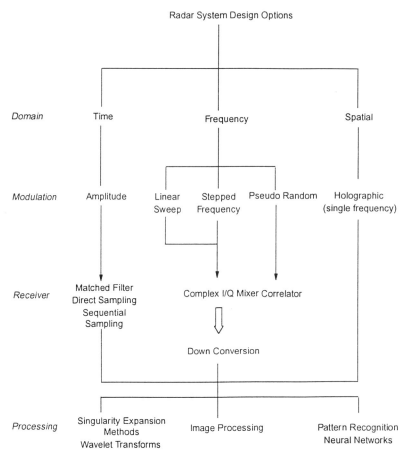

Figure 2.2   Surface-penetrating-radar-system design options

A block diagram of a generic radar system is shown in Figure 2.1. The source of energy can be an amplitude-frequency- or phase-modulated waveform and the selection of the bandwidth, repetition rate and mean power will depend upon the path loss and target dimensions. The transmit and receive antennas will usually be identical and will be selected to meet the characteristics of the generated waveform. The receiver must be suitable for the type of modulation and down conversion and possess an adequate dynamic range for the path losses which will be encountered. The various design options are shown in Figure 2.2 and will be discussed in Section 2.6 as well as in subsequent Chapters.

An initial estimate of the range performance of the radar can be gained by considering the following factors: path loss, target reflectivity, clutter and system dynamic range. The spatial resolution of the radar can be determined by considering the depth and plan resolution separately.

The majority of surface-penetrating-radar systems use an impulse time-domain waveform and receive the reflected signal in a sampling receiver. However, more use has been made of FMCW and stepped-frequency radar-modulation schemes in recent years and, as the cost of the components reduces, it may be expected that more of these systems will be used, as their dynamic range can be designed to be greater than that of the time-domain radar.

2.2 Range

2.2.1 Introduction

The range of a surface-penetrating radar is primarily governed by the total path loss, and the three main contributions to this are the material loss, the spreading loss and the target-reflection loss or scattering loss.

An example of a general method of estimation is given in this Section. Note that this contains many simplifying assumptions which later Chapters will discuss in more detail. The main assumption relates to the spreading loss. In conventional free-space radar, the target is in the far field of the antenna and spreading loss is proportional to the inverse fourth power of distance (R^{-4}). In many situations relating to surface-penetrating radar the target is in the near field and Fresnel zone and the (R^{-4}) relationship is no longer valid. However, for this example an R^{-4} spreading loss will be assumed.

The signal that is detected by the receiver undergoes various losses in its propagation path from the transmitter to the receiver. The total path loss for a particular distance is given by

$$L_t = L_e + L_m + L_{t1} + L_{t2} + L_s + L_a + L_{sc} \qquad (2.1)$$

where:

L_e = antenna efficiency loss, dB
L_m = antenna mismatch losses, dB

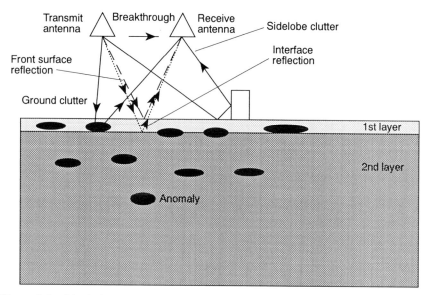

Figure 2.3   Physical layout of radar system

L_{t1} = transmission loss from air to material, dB
L_{t2} = retransmission loss from material to air, dB
L_s = antenna spreading losses, dB
L_a = attenuation loss of material, dB
L_{sc} = target scattering loss, dB.

At a frequency of, say, 1 GHz these losses may be estimated. In general, for accurate prediction, this calculation needs to be made over a wide band of frequencies but for this example a single frequency is assumed. The radar to be considered is an impulse radar using TEM horn antennas operated at a height of 1 m from the ground surface and the target is a planar interface at a depth of 0.2 m from the front surface of the material, as shown in Figure 2.3. For this example, the lateral dimensions of the planar interface can be considered to be infinitely large.

2.2.2 Antenna loss L_e

The antenna efficiency is a measure of the power available for radiation as a proportion of the power applied to the antenna terminals. For resistively loaded antennas, the efficiency is not high and is the result of the need for wideband operation. It would be expected that over an octave bandwidth the efficiency of a loaded antenna would be 4 dB lower than that of an unloaded antenna, which might have a loss of 2 dB. For other types of antenna, i.e. the short axial horn,

TEM horn etc., the antenna efficiency is higher and lower losses can be expected.

In the example under consideration it is assumed that

$$L_e = 2 \text{ dB per antenna, i.e. 4 dB for a pair of TEM horn antennas} \qquad (2.2)$$

2.2.3 Antenna-mismatch loss L_m

The antenna mismatch loss is a measure of how well the antenna is matched to the transmitter; usually little power is lost by reflection from antenna mismatch and L_M is in the order of 1 dB (or 2dB for both antennas).

2.2.4 Transmission-coupling loss L_{t1}

For TEM horn antennas operated at 1 m from the surface of the material, the transmission loss from the air to the material is given by

$$L_{t1} = 20 \log_{10} \left(\frac{2 Z_m}{Z_m + Z_a} \right) dB \qquad (2.3)$$

where:

$$
\begin{aligned}
Z_a &= \text{characteristic impedance of air} = 377 \ \Omega \\
Z_m &= \text{characteristic impedance of the material} \\
Z_m &= \left(\sqrt{\frac{\mu_0 \mu_r}{\epsilon_0 \epsilon_r}} \right) \frac{1}{(1 + \tan^2 \delta)^{\frac{1}{4}}} \left(\cos \frac{\delta}{2} + j \sin \frac{\delta}{2} \right)
\end{aligned} \qquad (2.4)
$$

Typically, for many earth materials,

$$Z_m = 100 \ \Omega$$

Hence

$$L_{t1} \simeq 8 \text{ dB}$$

2.2.5 Retransmission-coupling loss L_{t2}

The retransmission loss from the material to the air on the return journey is given by

$$L_{t2} = 20 \log \left(\frac{2 Z_a}{Z_a + Z_m} \right) \text{ dB} \qquad (2.5)$$

$$L_{t2} \simeq 4 \text{ dB}$$

2.2.6 Spreading loss L_s

The antenna spreading loss is conventionally related to the inverse fourth power of distance for a point reflector and in this example the ratio of the received power P_r to the transmitted power P_t is given by

$$\frac{P_r}{P_t} = \frac{G_t A_r}{(4\pi R^2)^2} \tag{2.6}$$

where:

$$G_t = \text{gain of transmitting antenna (TEM horn)} = 15$$
$$A_r = \text{receiving aperture (TEM horn)} = 4.10^{-2} \text{ m}^2$$
$$R = \text{range to the target} = 1.2 \text{ m.}$$

Hence L_S can be defined as

$$L_s = 10 \log_{10} \frac{G_t A_r}{(4\pi R^2)^2} \text{ dB}$$
$$L_s = 27 \text{ dB}$$

Note that the radar-range equation assumes a point-source scatterer which is not always the case. The range law may need adjusting for the different types of targets as shown in Table 2.1.

The nature of the target influences the magnitude of the received signal. The following approximate relationships apply for targets which extend across the zone illuminated by an antenna (i.e. its footprint). Note that the additional losses due to attenuation are ignored.

Considerably more backscattered energy will be returned from a planar reflector at a given depth compared with other target types exhibiting similar dielectric contrasts. As the target assumed in this example is a planar interface, a correction to the $1/R^4$ law is necessary, and taking this and the geometry into

Table 2.1   Adjustment of the range law for different types of target

Nature of target	Magnitude of received signal
Point scatterer (small void)	(target depth)$^{-4}$
Line reflector (pipeline)	(target depth)$^{-3}$
Planar reflector (smooth interface)	(target depth)$^{-2}$

account will reduce the spreading loss by approximately 7 dB. Hence $L_s = 20$ dB.

2.2.7 Target-scattering loss L_{SC}

For an interface between the material and a plane, where both the lateral dimensions of the interface and the overburden are large, then

$$L_{sc} = 20 \log \frac{Z_1 - Z_2}{Z_1 + Z_2} + 20 \log \sigma \qquad (2.7)$$

$Z_1 =$ characteristic impedance of the first layer of material

$Z_2 =$ characteristic impedance of second layer of material

$\sigma =$ target radar cross − section (as a proportion of intersection of both antenna beam patterns).

Typically L_{SC} would be in the order of 5 dB for the interface between the first and second layers. In this example, σ is considered to be unity, i.e. 0 dB, as the situation is equivalent to an infinite dielectric halfspace.

Where the physical dimensions of the interface or anomaly are small, the target-scattering loss L_{SC} increases owing to the geometry of the situation, and the returned signal becomes smaller. Under some conditions, the physical dimensions of the anomaly are such as to create a resonant structure which increases the level of the return signal and decreases the target-scattering loss. It is possible to distinguish air-filled voids and water-filled voids by examination of their resonant characteristics and the relative phase of the reflected wavelet. Water has a relative dielectric constant of 81 which will serve to reduce the resonant frequency of any void by a factor of 9 ($\sqrt{\epsilon_r}$), and this variation may be the means of identifying water-filled voids.

Table 2.2 gives an indication of the radar cross-sectional area in free space. The dimensions should be corrected for the different wavelengths within the material.

Many of the targets being searched for by subsurface-radar methods are nonmetallic, so their scattering cross-section is dependent on the properties of the surrounding dielectric medium. Where the relative permittivity of the target is lower than that of the surrounding medium, such as an air-filled void below a concrete ground slab, the interface does not produce a phase reversal of the backscattered wave. Conversely, when the scattering is caused by a metallic boundary or where the relative permittivity of the target is larger than the surrounding medium, phase reversal occurs in the backscattered wave. This phenomenon may be used as a way of distinguishing between conducting and nonconducting targets.

The physical shape of the target will influence the frequency and polarisation of the backscattered wave, and can be used as a means of preferential detection.

Table 2.2   Radar cross-sections

Scatterer	Aspect	Radar CSA	Symbols
Sphere		$\sigma = \pi a^2$	a = (radius)
Flat plate, arbitrary shape	Normal	$\sigma = \dfrac{4\pi A^2}{\lambda^2}$	A = (plate area)
Cylinder	Angle broadside	$\sigma = \dfrac{a\lambda \cos\theta \sin^2(kl \sin\theta)}{2\pi \sin^2\theta}$	a = (radius)
			l = (length)
Prolate spheroid	Axial	$\sigma = \dfrac{\pi b_0^4}{a_0^2}$	a_0 = (major axis)
			b_0 = (minor axis)
Triangular trihedral corner reflector	Symmetry axis	$\sigma = \dfrac{4\pi a^4}{3\lambda^2}$	l = (side length)

The effect of the high permittivity of typical soil means that some targets, such as thin-walled plastics pipes, produce a stronger radar return when buried than when in free space. In such circumstances the radar is responding primarily to the dielectric properties of the enclosed volume (i.e. the water or air-filled space within the pipe).

The type of target being sought (i.e. a sphere, a linear target such as a pipe or an interface) affects the choice of antenna type and configuration as well as the kind of signal-processing techniques which may be employed. Generally, parallel arrangements of dipole antennas are suitable for most targets whereas crossed dipoles are more appropriate for either small or linear targets.

2.2.8 Material-attenuation loss L_a

The attenuation loss of the material is given by

$$L_a = 8.686 R 2\pi f \sqrt{\frac{\mu_0 \mu_r \epsilon_0 \epsilon_r}{2}} \{\sqrt{(1 + \tan^2\delta)} - 1\} \ \mathrm{dB} \qquad (2.8)$$

where:

f = frequency, Hz

$\tan\delta$ = loss tangent of material

ϵ_r = relative permittivity of material

ϵ_0 = absolute permittivity of freespace

μ_r = relative magnetic susceptibility of material

μ_0 = absolute magnetic susceptibility of free space.

Table 2.3   Material loss at 100 MHz and 1GHz

Material	Loss at 100 MHz	Loss at 1 GHz
Clay (moist)	5–300 dB/m	50–3000 dB/m
Loamy soil (moist)	1–60 dB/m	10–600 dB/m
Sand (dry)	0.01–2 dB/m	0.1–20 dB/m
Ice	0.1–5 dB/m	1–50 dB/m
Fresh water	0.1 dB/m	1 dB/m
Sea water	1000 dB/m	10 000 dB/m
Concrete (dry)	0.5–2.5 dB/m	5–25 dB/m
Brick	0.3–2.0 dB/m	3–20 dB/m

Note that this loss should allow for a two-way travel; hence it is normal to double the one-way path loss.

A typical range of loss for various materials at 100 MHz and 1 GHz is shown in Table 2.3.

2.2.9 Total losses L_t

From the previous Sections the total losses that will occur at 1 GHz during transmission through 1 m of air and 0.2 m of material of 50 dB/m attenuation and then reflection from a material-air interface would be

$$L_t = 4\,\mathrm{dB} + 2\,\mathrm{dB} + 8\,\mathrm{dB} + 4\,\mathrm{dB} + 20\,\mathrm{dB} + 20\,\mathrm{dB} + 5\,\mathrm{dB} \qquad (2.9)$$
$$L_t = 63\,\mathrm{dB}$$

For a time-domain radar system it is more practical to consider peak voltages. The capability of a subsurface-radar system to detect a reflected signal of peak voltage V_r, if the peak transmitted voltage is V_T, can be termed the detectability D' of the radar system:

$$D' = 20\log\frac{V_t}{V_r}$$

The limiting factor of detectability is the noise performance of the receiver; hence the received voltage must be greater than the noise voltage generated by the latter.

For the previously calculated loss, if the transmitter generates a pulse of peak magnitude 50 V, the peak received signal would be 35 mV.

Most time-domain radar receivers can detect a 1 mV signal even without averaging; hence a reflected wavelet of peak amplitude of 35 mV should be capable of being detected. This is only the case if the clutter signals are also low,

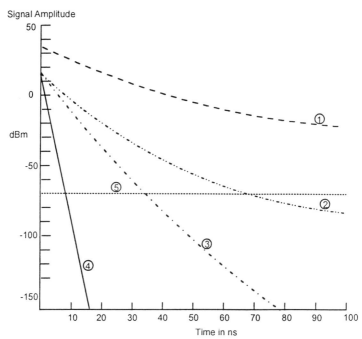

Figure 2.4   Signal amplitude against time

1 Spreading loss — — — —
2 Target loss — · · — · · —
3 Soil attenuation $\alpha = 30$dB/m — · — · —
4 Clutter profile ————
5 Receiver noise · · · · · · · ·

and the amplitude of the clutter signal should also be determined at the same range. Figure 2.4 shows a graph of the various signal levels plotted against range.

From the values of attenuation indicated above and the nature of the frequency dependence, it follows that, for a given signal-detection threshold, the maximum depth of investigation decreases rapidly with increasing frequency. Most subsurface-radar systems operate at frequencies less than 2 GHz. Figure 2.5 shows, for a range of materials measured by Cook (1975), the maximum depth of penetration at which radar is likely to be able to give useful information and the approximate upper frequency of operation. Typical maximum depths of penetration rarely exceed 20 wavelengths, except in very low-attenuation dielectric mediums. Often the depths of penetration will be much less, particularly in lossy dielectrics.

It might be thought that a surface-penetrating radar needs to have as low a frequency of operation as possible to achieve adequate penetration in wet materials. However, the ability to resolve the details of a target or

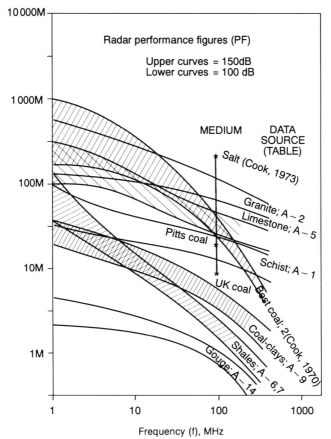

Figure 2.5   Radar probing range (after Cook)

separately detect two targets is proportional to the size or spacing of the target in relation to the wavelength of the incident radiation. Consequently, a high frequency is desirable for resolution. A compromise between penetration and resolution must be made and is an important consideration in the selection of either system bandwidth or the range of frequencies to be radiated.

Consideration needs also to be given to the fact that not only does attenuation reduce with frequency, but so does target-scattering cross-section. This leads to the situation where it is possible that, for certain targets, material properties and depths, the received signal reduces with frequency. This effect is shown in Figure 2.6 (Daniels, *et al.*, 1988) where the ratio between the reflected power, at frequencies of 50 MHz and 500 MHz, at the ground surface has been calculated for small-diameter metallic cylinders (broken line) or nonmetallic (solid line).

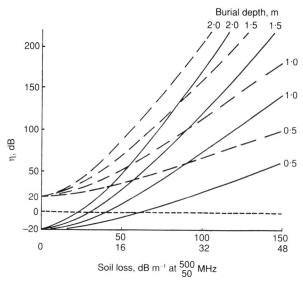

Figure 2.6   Received signal level of various targets (after Daniels)

2.2.10 Velocity of propagation

It can easily be recognised that, if the propagation velocity can be measured, or derived, an absolute measurement of depth or thickness can be made. For homogeneous isotropic materials, the relative propagation velocity v_r can be calculated from

$$v_r = \frac{c}{\sqrt{\epsilon_r}} \text{ metres per second} \tag{2.10}$$

and the depth derived from

$$d = v_r \frac{t_r}{2} \text{ metres} \tag{2.11}$$

where

ϵ_r is the relative permittivity

t_r is the transit time to and from the target.

In most practical trial situations, the relative permittivity will be unknown. The velocity of propagation must either be measured *in-situ*, estimated by means of direct measurement of the depth to a physical interface or target (ie by trial holing), or by calculation by means of multiple measurements. From Figure 2.7 it can be seen that, if a hyperbolic spreading function can be measured, then the propagation velocity can be derived from

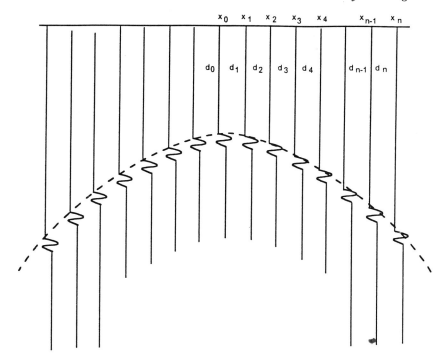

Figure 2.7   Hyperbolic spreading function

$$v_r = 2 \sqrt{\frac{x_{n-1}^2 - x_n^2}{t_{n-1}^2 - t_n^2}} \tag{2.12}$$

and the depth to the target

$$d_0 = \frac{v_r t_0}{2}$$

An alternative method of calculating the depth of a single planar reflector is by means of the common-depth-point method. If both transmitting and receiving antennas are moved equal distances from the common centre point, the same apparent reflection position will be maintained.

The depth of the planar reflector can be derived from

$$d = \sqrt{\frac{x_{n-1}^2 t_n^2 - x_n^2 t_{n-1}^2}{t_{n-1}^2 - t_n^2}} \tag{2.13}$$

where the two positions of the antenna are shown in Figure 2.8.

The variation of permittivity with frequency in wet dielectrics implies that

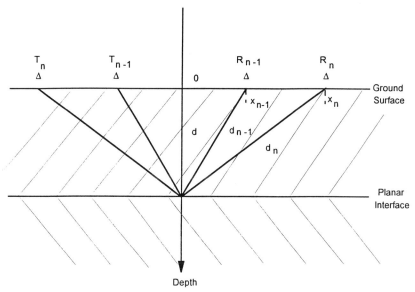

Figure 2.8   Common-depth-point depth estimation

there will be some variation in the velocity of propagation with frequency. The magnitude of this effect will generally be small for the range of frequencies typically employed for subsurface-radar work. A dielectric exhibiting this characteristic is said to be 'dispersive'. Where the material has different propagation characteristics in different directions, it is said to be anistropic and an example is coal in the seams prior to excavation, where the propagation characteristics normal to the bedding plane are different from those parallel to the plane.

In free space the propagation velocity c is $3 \times 10^8 m/s$. The velocity in air is very similar to that in free space and is normally taken as the same. In subsurface-radar work the elapsed time between the transmitted and received pulses is measured in nanoseconds $(10^{-9}$ s$)$ because of the short travel-path lengths involved.

Propagation velocity reduces with increasing relative permittivity. The

Table 2.4   Material propagation characteristics

Material	Relative permittivity ϵ_r	Propagation velocity	Wavelength λ	
			100 MHz	1 GHz
		cm/ns	cm	cm
Air	1	30	300	30
Concrete	9	10	100	10
Freshwater	80	3.35	33	3

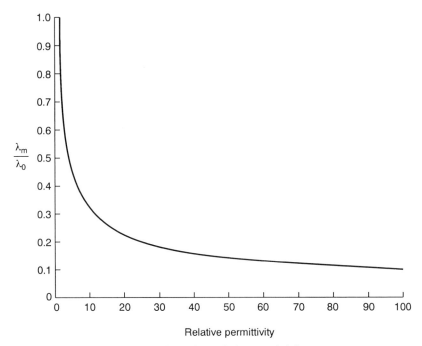

Figure 2.9   Normalised wavelength against relative permittivity

wavelength within the material also decreases as the velocity of propagation slows in accordance with the relationship

$$\lambda_m = \frac{v_r}{f} \text{ metres} \qquad (2.14)$$

The result of these effects is illustrated by Table 2.4.

The reduction in wavelength in earth materials assists in the accuracy of measurement of depth by improving the resolution compared with that in air. The relationship between wavelength and relative permittivity is shown graphically in Figure 2.9. Wetter materials have larger permittivities and hence lower propagation velocities and shorter wavelengths than dry ones. Most materials contain moisture, usually with some measure of conducting salts. If the water content is large enough, the permittivity of the material may be determined primarily by the dielectric properties of water and only secondarily by those of the (dry) material.

2.3 Clutter

The clutter that affects a surface penetrating radar can be defined as those signals that are unrelated to the target scattering characteristics but occur in the same sample-time window and have similar spectral characteristics to the target wavelet. This is a somewhat different definition from conventional radar clutter and should be borne in mind when considering conventional methods of clutter filtering such as MTI which would be inappropriate to apply to surface-penetrating-radar data.

Clutter can be caused by breakthrough between the transmit and receive antennas as well as multiple reflections between the antenna and the ground surface. Clutter will vary according to the type of antenna configuration, and the parallel-planar-dipole arrangement is one where the stability of the level of breakthrough is most constant. Typically, a maximum level of −40 dB to −50 dB is encountered.

The planar crossed-dipole antenna can be designed and manufactured to provide very low levels of breakthrough (> 70 dB). However, it then becomes very susceptible to 'bridging' by dielectric anomalies on the near surface which can degrade the breakthrough in a random manner as the antenna is moved over the ground surface. The variability of the breakthrough is unfortunate, as it is not usually amenable to signal processing. The same problem is encountered with planar spiral antennas. Local variations in the characteristic impedance of the ground can also cause clutter, as can inclusions of groups of small reflection sources within the material. In addition, reflections from targets in the sidelobes of the antenna, often above the ground surface, can be particularly troublesome. This problem can be overcome by careful antenna design and by incorporating radar-absorbing material to attenuate the side- and back-lobe radiation from the antenna.

In general, clutter is more significant at short range times and reduces at longer times.

It is possible to quantify the rate of change of the peak clutter signal level as a function of time, as in many cases this parameter sets a limit to the detection capability of the radar system. The effect of clutter on system performance is shown in Figure 2.4 which illustrates the consequent limitation on near-range radar performance.

Various techniques have been investigated in the search for a method of reducing clutter and, for impulse radars using TEM horns or FMCW radars using ridged horns and reflectors, it has been found possible to angle the boresight of the horn antennas to take advantage of the critical angle, thereby suppressing to some extent the ground-surface reflection.

2.4 Depth resolution

There are some applications of subsurface radar, such as road-layer-thickness measurement, where the feature of interest is a single interface. Under such circumstances it is possible to determine the depth sufficiently accurately by measuring the elapsed time between the leading edge of the received wavelet and a reference time such as the front-surface reflection provided that the propagation velocity is accurately known.

However, when a number of features may be present, such as in the detection of buried pipes and cables, then a signal having a larger bandwidth is required to be able to distinguish between the various targets and to show the detailed structure of a target. In this context it is the bandwidth of the received signal which is important, rather than that of the transmitted wavelet. The 'earth' acts as a lowpass filter which modifies the transmitted spectrum in accordance with the electrical properties of the propagating medium. The results from a simplified model of this situation are shown in Figure 2.10 where the general pulse stretching can be seen for different rates of attenuation.

The required receiver bandwidth B' can be determined by considering the power spectrum of the received signal. The power spectrum results from the Fourier transform of the received signal wavelet. If the envelope of the wavelet function is considered, it is possible from the Rayleigh criterion for resolution to determine a receiver bandwidth. An alternative definition of the receiver bandwidth is given by Cook (1975) and is derived from the autocorrelation function of the signal wavelet.

If $f(t)$ is the wavelet, then the autocorrelation function is given by

$$R_{11}(\tau) = \int_{-\infty}^{\infty} f_1(t)f(t - \tau)\mathrm{d}t \tag{2.15}$$

The general shape of the autocorrelation function shown in Figure 2.11, which is related to the matched filtering resolution, can be used to define the bandwidth requirement. The autocorrelation function is, of course, related to the power spectrum of the received waveform and is therefore a useful measure. A receiver bandwidth in excess of 500 MHz, and typically 1 GHz, is required to provide a typical resolution of between 5 and 20 cm, depending on the relative permittivity of the material.

Although greater depth resolution is achieved in wetter materials for a given transmitted bandwidth, earth materials with significant water content tend to have higher attenuation properties. This characteristic reduces the effective bandwidth, tending to balance out the change so that, within certain bounds, the resolution is approximately independent of loss within the propagating material.

Where interfaces are spaced more closely than one half wavelength, the reflected signal from one interface will become convolved with that from the other, as shown in Figure 2.12. In such circumstances, some form of deconvolution processing would be required in order to recognise the responses from the individual interfaces and to enable them to be characterised and traced. However, such processing is not normally carried out during standard commercial radar surveys.

2.5 Plan resolution

The plan resolution of a subsurface-radar system is important when localised targets are sought and when there is a need to distinguish between more than

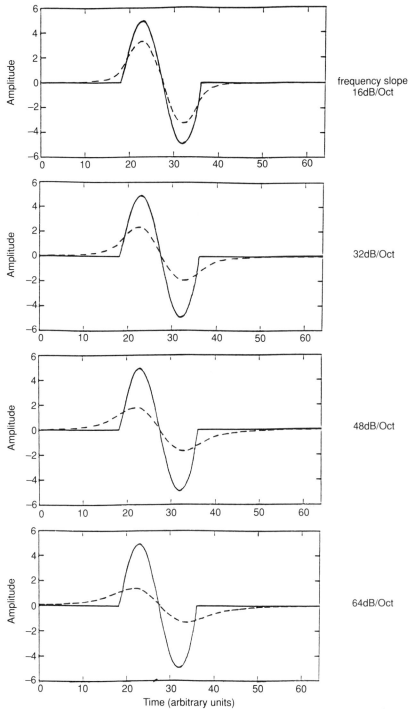

Figure 2.10   Effect of ground attenuation on pulse length

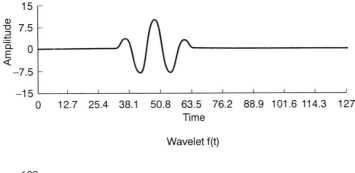

Wavelet f(t)

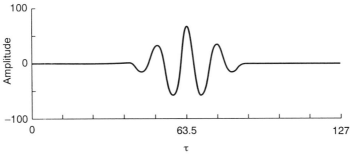

Aurocorrelation function of f(t)

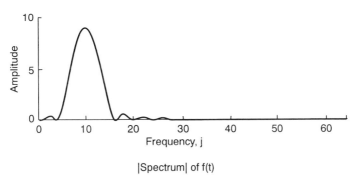

|Spectrum| of f(t)

Figure 2.11   Autocorrelation of wavelet function

one at the same depth. Where the requirement is for location accuracy, which is primarily a topographic-surveying function, the system requirement is less demanding.

The plan resolution is defined by the characteristics of the antenna and the signal processing employed. In general, to achieve an acceptable plan resolution requires a high-gain antenna. This necessitates an antenna with a significant aperture at the lowest frequency transmitted. To achieve small antenna dimensions and high gain therefore requires the use of a high carrier frequency,

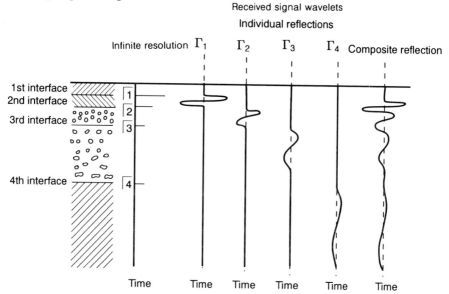

Figure 2.12   Convolution of multiple interface reflections

which may not penetrate the material to sufficient depth. When choosing equipment for a particular application it is necessary to compromise between plan resolution, size of antenna, the scope for signal processing and the ability to penetrate the material.

Plan resolution improves as attenuation increases, assuming that there is sufficient signal to discriminate under the prevailing clutter conditions. In low-attenuation media, the resolution obtained by the horizontal-scanning technique is degraded, but under these conditions the use of advanced signal-processing techniques becomes feasible. These techniques typically require measurements made using transmitter and receiver pairs at a number of antenna positions to generate a synthetic aperture or focus the image. Unlike conventional radars, which generally use a single antenna, most surface-penetrating-radar systems use separate transmit and receive antennas in what has been termed a bistatic mode. However, as the antenna configuration is normally mobile, the term bistatic is not really relevant.

The descriptions normally applied to the modes of geophysical survey appear more relevant and are therefore introduced. Geophysicists classify surveys in four main modes: common source, common receiver, common offset and common depth point, as shown in Figure 2.13.

Most surface-penetrating-radar surveys use a common-offset survey mode in which the separation between the transmitter and receiver is fixed. However, both common-depth-point and common-source or receiver modes have also been used, but require different signal-processing approaches.

In the common-offset mode the transmitter and receiver antennas are scanned above the ground surface over a buried target, as shown in Figure 2.13. The

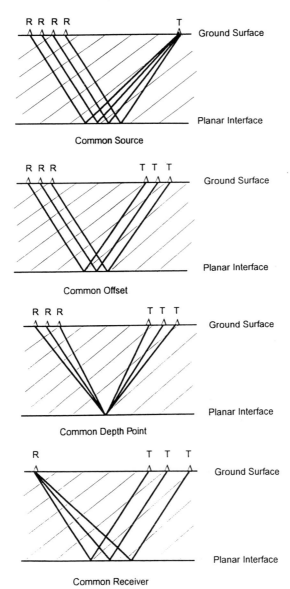

Figure 2.13 Geophysical survey modes

received power for a point source scatterer in the far field can be shown to be proportional to

$$P_r = \frac{P_t \cos^4 \theta}{d^4} \exp(-2\alpha d \sec \theta)$$

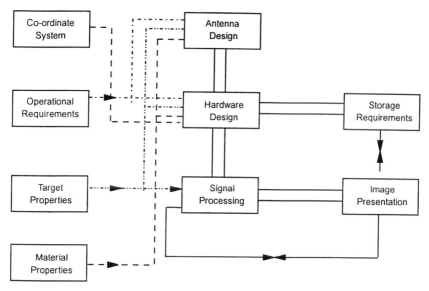

Figure 2.14   System-design considerations

where α is the attenuation coefficient and θ is the angle between the midpoint of the combined transmit-receive antenna and the vertical to the target.

Where the plan resolution is defined as the half-power points of the spatial response of the scatterer at the plane of the surface, the resolution is approximated by

$$\Delta x = 4d \sqrt{\frac{\ln 2}{2 + \alpha d}}$$

This approximation takes no account of the antenna-beam pattern in either x or y directions. However, it does indicate that, as the attenuation increases, the plan resolution improves, provided that adequate signal-to-noise and signal-to-clutter ratios are maintained. Note that, in low-attenuation materials, synthetic aperture processing can be applied and plan resolution is recovered.

Typically, the improvement in resolution is most noticeable at depths greater than 1 m and an improvement in resolution of 30% would be found in the plan resolution of targets buried at 2 m in materials of 9 dB/m and 30 dB/m attenuation.

2.6 System considerations

The majority of surface-penetrating radars are based on the time-domain impulse design. Alternative design options can be considered and experimental versions of stepped-frequency or synthesised-impulse, as well as frequency-modulated (FMCW), have been designed and built.

While the different modulation techniques are considered in detail in Chapter 5 it is useful to summarise the general attributes of each option.

Time-domain impulse-radar systems are available commercially and manufacturers usually offer a range of antennas and pulse lengths to suit the desired probing range. Depths of greater than 30 m require pulse lengths of the order of 40 ns (approximately a bandwidth of 50 MHz at a centre frequency of 25 MHz) and very short-range precision probing may use pulse lengths of the order of or less than 1.0 ns, that is an approximate bandwidth of 2 GHz at a centre frequency of 1 GHz.

Planar impulse-radar antennas generally operate closely coupled to the ground and are usually designed so that the polarisations of the transmitted and received signals are parallel. The exception to this is the crossed-dipole antenna which has been used for detecting either linear features such as pipes, cables and cracks in the material or small targets such as buried plastics mines.

Most antennas have relatively small footprints which means that rapid and wide-area surveying can only be achieved with multichannel radar systems. For road survey, such methods are cost effective and practical. An alternative to the planar antenna is the TEM horn which can be used with a surface-to-antenna spacing of up to 1 m.

Although alternative modulation methods to the impulse radar have been used, there are very few commercially available FMCW, stepped-frequency or synthesised pulse-radar equipment, although this situation could change in the future.

Whatever system is considered, it is important to consider the receiver dynamic range and sensitivity rather than the ratio of the peak transmitted signal to the minimum detectable signal.

Figure 2.14 illustrates diagrammatically the various parameters that influence the selection of a radar system and its characteristics.

A practical example will illustrate the reason for this. If a surface-penetrating radar is being used to survey the road and traverses a metal cover on the surface of the road, the received signal will be maximum and may well saturate the receiver circuits. If the receiver cannot recover from this high-level input within a few nanoseconds, all low-level signals caused by targets deeper than and adjacent to the cover will go undetected. It is therefore the receiver performance and the method of signal downconversion which must be considered as defining the overall performance. A receiver with a true dynamic range of 60 dB followed by an analogue-to-digital converter with a dynamic range of 96 dB (16 bits) is only capable of providing an effective dynamic range of 60 dB.

Great care should be taken when interpreting specifications to ascertain the true system performance.

2.7 References

COOK, J. (1975): Radar transparencies of mine and tunnel rocks. *Geophysics*, **40**, 865–885
DANIELS, D. J., GUNTON, D. J., and SCOTT, H. F. (1988): Introduction to subsurface radar. *IEE Proc.* F, **135**, (4), 278-321

Chapter 3

Properties of materials

3.1 Introduction

A significant number of researchers have investigated extensively the dielectric properties of earth materials. They have shown experimentally that, for most materials which constitute the shallow subsurface of the earth, which in this case is taken to be a zone of depths of 100 m or less, the attenuation of electromagnetic radiation rises with frequency and that, at a given frequency, wet materials exhibit higher losses than dry ones. From this generalisation a number of predictions can be made relating to the performance of a surface-penetrating-radar system. Before this can be done, it is necessary to understand those characteristics of materials which affect both the velocity of propagation and attenuation.

An order-of-magnitude indication of the basic dielectric characteristics of various materials can be gauged from Table 3.1 which shows their conductivity and relative permittivity measured at a frequency of 100 MHz. The velocity of propagation is primarily governed by the relative permittivity of a material, which depends primarily upon its water content. At low microwave frequencies, including the range over which surface-penetrating radar systems operate, water has a relative permittivity ϵ_r of approximately 80, while the solid constituents of most soils and man-made materials have, when dry, a relative permittivity in the range 2–9. The measured values of ϵ_r for soils and building materials lie mainly in the range 4 to 40. The absolute permittivity also varies with frequency, but is generally sensibly constant for most materials over the range of frequencies utilised for surface-penetrating-radar work. The attenuation of a material is a more complex relationship and will be discussed in more detail in subsequent Sections of this Chapter.

The physical models which are used to predict the propagation of electromagnetic waves in dielectric materials have two main sources: electromagnetic-wave theory and geometrical optics. The latter method is only relevant when the wavelength of the electromagnetic radiation employed is considerably shorter than the dimensions of the object or medium being illuminated and when the materials involved can be considered to be electrical insulators. Optical theory is therefore most relevant to dry materials. Materials containing appreciable amounts of moisture will behave as conducting dielectrics, especially if the water contains ions. Most naturally occurring waters have some degree of ionic conduction and so act as aqueous electrolytes.

Table 3.1 Typical range of dielectric characteristics of various materials measured at 100 MHz (reported)

Material	Conductivity	Relative permittivity ϵ_r
	S/m	F/m
Air	0	1
Asphalt: dry	10^{-3}–10^{-2}	2–4
Asphalt: wet	10^{-2}–10^{-1}	6–12
Clay: dry	10^{-3}–10^{-1}	2–6
Clay: saturated	10^{-1}–1	15–40
Coal: dry	10^{-2}	3.5
Coal: wet	10^{-1}	8
Concrete: dry	10^{-3}–10^{-2}	4–10
Concrete: wet	10^{-2}–10^{-1}	10–20
Freshwater	10^{-4}–10^{-2}	81
Freshwater ice	10^{-3}	4
Granite: dry	10^{-8}–10^{-6}	5
Granite: wet	10^{-3}–10^{-2}	7
Limestone: dry	10^{-9}–10^{-6}	7
Limestone: wet	10^{-2}–10^{-1}	8
Permafrost	10^{-5}–10^{-2}	4–8
Rock salt: dry	10^{-4}	4–7
Sand: dry	10^{-7}–10^{-3}	4–6
Sand: saturated	10^{-4}–10^{-2}	10–30
Sandstone: dry	10^{-9}–10^{-6}	2–3
Sandstone: wet	10^{-5}–10^{-6}	5–10
Seawater	4	81
Seawater ice	10^{-2}–10^{-1}	4–8
Shale: saturated	10^{-2}–10^{-1}	6–9
Snow: firm	10^{-6}–10^{-5}	8–12
Soil: sandy dry	10^{-4}–10^{-2}	4–6
Soil: sandy wet	10^{-2}–10^{-1}	15–30
Soil: loamy dry	10^{-4}–10^{-3}	4–6
Soil: loamy wet	10^{-2}–10^{-1}	10–20
Soil: clayey dry	10^{-4}–10^{-1}	4–6
Soil: clayey wet	10^{-1}–1	10–15

The variability of both material parameters and local geological conditions that is encountered in real life is such as to cause great difficulty in accurate prediction of propagation behaviour. This point should be noted when assessing the value of predictive methods, as an accurate description by means of a theoretical approach may not adequately describe a situation with many degrees of freedom. Similarly, it is often difficult to replicate the bulk material electromagnetic characteristics of a material in laboratory conditions using test cells. Even if the moisture content is correctly replicated, differences in density between *in-situ* and laboratory samples are difficult to minimise.

Although all ultra-wideband radars transmit energy over at least an octave frequency band, it is possible to make an order-of-magnitude estimation of their performance based on the centre frequency of operation. The objective of this Chapter is to provide an introduction into those parameters of materials which affect the performance of surface-penetrating radars. The relatively standard treatment given in this Chapter will, of course, only provide a guide to system performance and it is possible to consider a more complex treatment. However, the experimental practitioner will soon realise that accurate modelling is difficult to achieve within any reasonable budget owing to the number of the degrees of freedom needed. For this reason, a simplified treatment is considered an adequate introduction.

3.2 Propagation of electromagnetic waves in dielectric materials

Maxwell's equations are the foundation for the consideration of the propagation of electromagnetic waves. In a perfect dielectric material the magnetic susceptibility μ and electric permittivity ϵ are constants, that is they are independent of frequency and the medium is not dispersive. In a perfect dielectric no propagation losses are encountered and hence there is no consideration of the attenuation which occurs in real dielectric media.

Plane waves are good approximations to real waves in many practical situations, particularly in low-loss and resistive media such as dry limestone and sands. More complicated electromagnetic-wave fronts can be considered as a superimposition of plane waves, and this method may be used to gain an insight into more complex situations.

As a starting point, electromagnetic-wave propagation can be represented by a one-dimensional wave equation of the following form. Propagation is taken along the z axis, with perpendicular electric (E) and magnetic (H) fields as shown in Figure 3.1.

$$\frac{\partial^2 E}{\partial z^2} = \mu\epsilon \frac{\partial^2 E}{\partial t^2} \tag{3.1}$$

where the velocity of propagation

$$v = \frac{1}{\sqrt{(\mu\epsilon)}} \text{ metres per second} \tag{3.2}$$

the velocity of light in free space is

$$c = \frac{1}{\sqrt{(\mu_0\epsilon_0)}} = 3 \times 10^8 \text{ m/s} \tag{3.3}$$

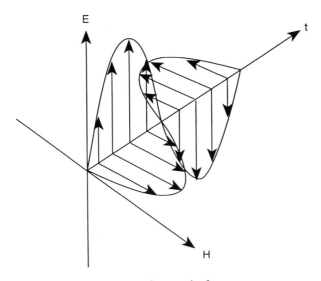

Figure 3.1 Propagation of electromagnetic wave in free space

where:

absolute magnetic permeability of free space $\mu_0 = 1.26 \times 10^{-6}\text{H/m}$
absolute electric permittivity of free space, $\epsilon_0 = 8.84 \times 10^{-12}\text{F/m}$
absolute magnetic permeability of medium, $\mu = \mu_0\mu_r$
absolute electric permittivity of medium, $\epsilon = \epsilon_0\epsilon_r$

and

ϵ_r = relative permittivity, having a value in the range 1 to $\sim$ 80 for most geological materials
μ_r = relative magnetic permeability, being 1 for nonmagnetic geologic materials.

Hence

$$v_r = \frac{c}{\sqrt{\epsilon_r}} \text{ m/s} \tag{3.4}$$

The intrinsic impedance (the ratio of the electric to the magnetic field) of the medium

$$\eta = \sqrt{\left[\frac{\mu}{\epsilon}\right]} \text{ ohms} \tag{3.5}$$

A wave propagating in the positive z direction in a perfect dielectric can be described by the equation

$$E(z) = E_0 \, e^{-jkz} \qquad (3.6)$$

where the phase constant

$$k = \frac{\omega}{v} = \omega\sqrt{(\mu\epsilon)} \text{ per metre} \qquad (3.7)$$

This describes the change in phase per unit length for each wave component; it may be considered as a constant of the medium for a particular frequency and is then known as the wave number. It may also be referred to as the propagation factor for the medium.

The wavelength λ is defined as the distance the wave propagates in one period of oscillation. It is then the value of z which causes the phase factor to change by 2π:

$$k\lambda = 2\pi \qquad (3.8)$$

rearranging

$$\lambda = \frac{2\pi}{\omega\sqrt{(\mu\epsilon)}} = \frac{v}{f} \text{ metres} \qquad (3.9)$$

This is the common relationship between wavelength, phase velocity and frequency.

In optics it is common to utilise a refractive index n given by

$$n = \frac{c}{v} = \sqrt{(\mu_r \epsilon_r)} \qquad (3.10)$$

taking $\mu_r = 1$, $n = \sqrt{\epsilon_r}$ for the frequency being considered.

Electromagnetic waves propagating through natural media experience losses to either the electric (E) or magnetic (H) fields, or both. This causes attenuation of the original electromagnetic wave. For most materials of interest in surface-penetrating radar, the magnetic response is weak and need not be considered as a complex quantity; unlike the permittivity and conductivity. For lossy dielectric materials, absorption of electromagnetic radiation is caused by both conduction and dielectric effects. It is not possible to distinguish, by measurement at a single frequency, the separate components of loss for such materials.

In general, the complex permittivity ϵ and the complex conductivity σ may be expressed as:

$$\epsilon = \epsilon' - j\epsilon'' \qquad (3.11)$$

and

$$\sigma = \sigma' - j\sigma'' \qquad (3.12)$$

where

$$\begin{aligned} \epsilon' \text{ and } \sigma' &= \text{ real parts} \\ \epsilon'' \text{ and } \sigma'' &= \text{ imaginary parts} \\ j &= \sqrt{-1} \end{aligned}$$

The nature of the parameter ϵ' relates to the electric permittivity, which may also be expressed in terms of relative permittivity. The parameter ϵ'' relates to losses associated with both conductivity and frequency. For practical purposes, at frequencies up to 1 GHz and conductivities below 0.1 S/m, the effect of the ϵ'' term will be small and is commonly disregarded (i.e. ϵ taken as the real components, ϵ') in such circumstances.

When measurements are made on a conducting dielectric, the parameters measured are the apparent permittivity $\tilde{\epsilon}$ and apparent conductivity $\tilde{\sigma}$:

$$\tilde{\epsilon} = \epsilon'_e - j\epsilon''_e \qquad (3.13)$$

$$\tilde{\sigma} = \sigma'_e - j\sigma''_e \qquad (3.14)$$

The behaviour of a material may be specified either by its apparent permittivity or, equivalently, by its apparent conductivity since

$$\tilde{\sigma} = j\omega\tilde{\epsilon} \qquad (3.15)$$

In terms of wave-propagation equations, ϵ and σ always occur in the combination

$$\sigma + j\omega\epsilon \equiv \sigma'_e + j\omega\epsilon'_e \qquad (3.16)$$

where

$$\begin{aligned} \sigma'_e &= \text{ the real effective conductivity} \\ \epsilon'_e &= \text{ the real effective permittivity} \end{aligned}$$

$$\sigma'_e = \sigma' + \omega\epsilon'' \qquad (3.17)$$

$$\epsilon'_e = \epsilon' - \frac{\sigma''}{\omega} \qquad (3.18)$$

From eqn. 3.6 the propagation of an electromagnetic field E_0 originating at

$z = 0$, $t = 0$ in a conducting dielectric can be described by $E(z, t)$ at a distance z and time t by

$$E(z, t) = E_0 \cdot e^{-\alpha z} \cdot e^{j(\omega t - \beta z)} \tag{3.19}$$

the first exponential function is the attenuation term and the second the propagation term.

From the first exponential function, it is seen that at a distance $z = 1/\alpha$ the attention is $1/e$. This distance is known as the skin depth and provides a useful guide to the useful penetration depth of a surface-penetrating radar system. However, there are a number of other factors which influence the effective penetration depth, notably the strength of reflection from the target sought and the degree of clutter suppression of which the system is capable. These may reduce the calculated performance and must also be considered.

In general, the parameters of interest for subsurface-radar applications are the attenuation and velocity of wave propagation.

In a conducting dielectric the phase coefficient is complex and is

$$k = \omega \sqrt{\mu(\epsilon' - j\epsilon'')} \tag{3.20}$$

The wave number may be separated into real and imaginary parts:

$$jk = \alpha + j\beta = j\omega \sqrt{\mu \epsilon' \left(1 - j\frac{\epsilon''}{\epsilon'}\right)} \tag{3.21}$$

where

$$\alpha = \text{ attenuation factor}$$
$$\beta = \text{ phase coefficient.}$$

The parameters α and β can be related to σ and $j\omega\epsilon$ and give expressions for α and β as shown below:

$$\alpha = \omega \left[\frac{\mu \epsilon'}{2} \left(\sqrt{1 + \left(\frac{\epsilon''}{\epsilon'}\right)^2} - 1\right)\right]^{\frac{1}{2}} \tag{3.22}$$

$$\beta = \omega \left[\frac{\mu \epsilon'}{2} \left(\sqrt{1 + \left(\frac{\epsilon''}{\epsilon'}\right)^2} + 1\right)\right]^{\frac{1}{2}} \tag{3.23}$$

The dimensionless factor ϵ_e''/ϵ_e' is more commonly termed the material loss tangent

$$\frac{\epsilon_e''}{\epsilon_e'} = \tan \delta \qquad (3.24)$$

The above expressions may be rearranged to provide the attenuation coefficient α' in decibels per metre and the wave velocity v in metres per second:

$$\alpha' = 8.66\alpha \ \mathrm{dB/m} \qquad (3.25)$$

It can be seen from the above expressions that the attenuation coefficient of a material is, to a first order, linearly related (in dB/m) to frequency. It is not sufficient to consider only the low-frequency conductivity σ_0 when attempting to determine the loss tangent over the frequency range 1×10^{-7} to 1×10^{10} Hz. For a material which is dry and relatively lossless, it may be reasonable to consider that $\tan \delta$ is constant over that frequency range. However, for materials which are wet and lossy such an approximation is invalid, as

$$\tan \delta = \frac{\sigma' + \omega\epsilon''}{\omega\epsilon' - \sigma''} \qquad (3.26)$$

It is important to consider the magnitude of σ'' and ϵ'' in attempting to determine the value of $\tan \delta$. In general, for lossy earth materials $\tan \delta$ is large at low frequencies, exhibits a minimum at around 1×10^8 Hz and increases to a maximum at several gigahertz, remaining constant thereafter.

Accurate determination of attenuation must therefore consider all the coefficients which comprise the expression for $\tan \delta$.

An approximation which enables an order-of-magnitude indication is that when σ is small

$$\tan \delta \simeq \frac{\sigma'}{\omega\epsilon'} \qquad (3.27)$$

It should be noted that the complex dielectric coefficient, and hence the loss factor, of a soil is affected by both temperature and water content. The general effect of increasing the temperature is to reduce the frequency of the dielectric relaxation, while increasing the water content also increases the value of the loss factor while shifting its peak frequency down.

It is observed the frequency of the maximum dielectric loss of the water relaxation in soils is reduced and occurs over a more limited frequency range when compared with conductive water.

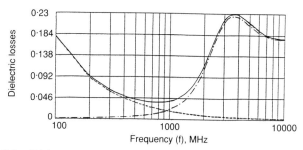

Figure 3.2   Dielectric losses as a function of frequency for a medium-loss soil

Tan δ can increase with frequency over the range 1×10^8 to 1×10^{10} Hz as the dipolar losses associated with the water content of the material become more significant and the conductivity losses reduce. Hence

$$\tan \delta = \text{ conductivity losses } + \text{ dipolar losses}$$

$$\tan \delta = \frac{\sigma_{dc}}{\omega \epsilon_0 \epsilon_r} + \text{ dipolar losses} \qquad (3.28)$$

It is necessary to determine the dipolar losses from a consideration of the dielectric relaxation spectrum of the soil/water under consideration. Typically for silty clay soils, the conductivity losses could be 0.5×10^{-1} at $f = 10^8$ Hz while the dipolar losses could increase from a similar value to 10 at $f = 5 \times 10^9$ Hz. A graph of dielectric losses for a medium loss soil is shown in Figure 3.2.

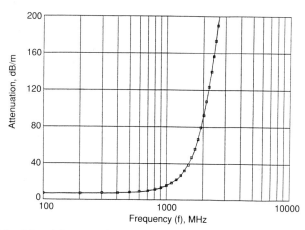

Figure 3.3   Material attenuation as a function of frequency for a medium-loss soil

A graph of attenuation against frequency for such a material is shown in Figure 3.3.

The velocity of propagation is also slowed by an increase of loss tangent as well as relative dielectric coefficient, as

$$v = c \left[\frac{\epsilon'_e}{2\epsilon_0} \left(\sqrt{(1 + \tan^2 \delta)} + 1 \right) \right]^{\frac{1}{2}} \tag{3.29}$$

However, $\tan \delta$ must be significantly greater than 1 for any slowing to occur and it is reasonable to assume that, for

$$\tan \delta < 1$$
$$v = \frac{c}{\sqrt{\epsilon_r}} \tag{3.30}$$

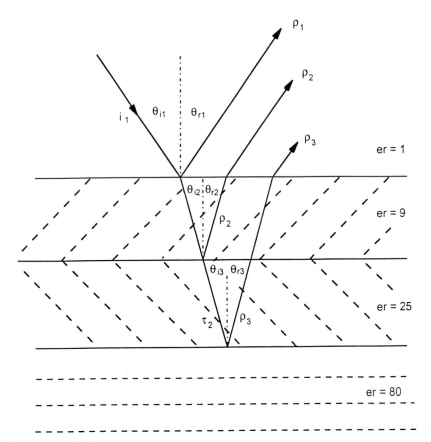

Figure 3.4   Multilayer dielectric

In any estimation of received signal level it is necessary to consider the coefficients of reflection and transmission as the wave passes through the dielectric to the target, as shown in Figure 3.4. To do this we need to consider the intrinsic impedance of the various materials.

The intrinsic impedance η of a medium is the relationship between the electric field E and the magnetic field H.

$$\eta = \frac{E}{H} \qquad (3.31)$$

η is a complex quantity which is calculated according to

$$\eta = \left[\frac{-j\omega\mu}{\sigma - j\omega\epsilon}\right]^{\frac{1}{2}} \qquad (3.32)$$

The intrinsic impedance of the medium becomes

$$\eta = \left(\frac{\mu}{\epsilon}\right)^{\frac{1}{2}} = \sqrt{\mu\left(\epsilon'\left(1 - j\frac{\epsilon''}{\epsilon'}\right)\right)^{-\frac{1}{2}}} \text{ ohms} \qquad (3.33)$$

At the boundary between two media, some energy will be reflected and the remainder transmitted. The reflected field strength is described by the reflection coefficient r:

$$r = \frac{\eta_2 - \eta_1}{\eta_2 + \eta_1} \qquad (3.34)$$

where η_1 and η_2 are the impedances of media 1 and 2, respectively.

In a nonconducting medium, such as dry soil or dry concrete, and when considering only a single frequency of radiation, the above expression may be simplified and rewritten as

$$r = \frac{\sqrt{\epsilon_{r2}} - \sqrt{\epsilon_{r1}}}{\sqrt{\epsilon_{r2}} + \sqrt{\epsilon_{r1}}} \qquad (3.35)$$

where ϵ_r is the relative permittivity of the medium.

The reflection coefficient has a positive value when $\epsilon_{r2} > \epsilon_{r1}$, such as where an air-filled void exists in a dielectric material. The effect on a pulse waveform is to change the phase of the reflected wavelet so that targets with different relative permittivities to the host material show different phase patterns of the reflected signal. However, the amplitude of the reflected signal is affected by the propagation dielectric of the host material, the geometric characteristics of the target and its dielectric parameters. Figures 3.5 and 3.6 show the predicted

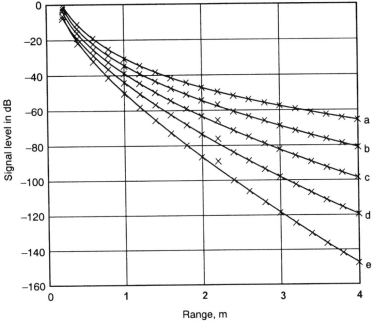

Figure 3.5 *Received signal level against range at 100 MHz*
  a Attenuation = 3.7 dB/m, loss tangent = 0.2
  b Attenuation = 7.5 dB/m, loss tangent = 0.42
  c Attenuation = 11.8 dB/m, loss tangent = 0.68
  d Attenuation = 16 dB/m, loss tangent = 1.03
  e Attenuation = 23.75 dB/m, loss tangent = 1.558

relative received signal at frequencies of 100 MHz and 1 GHz for a target of
0.1 m² cross-section over a range of depth of cover up to 4 m.

The previous description is an elementary description of a situation, which is
described with considerably more precision by Wait (1960) and King and Wu
(1993).

Wait initially considers the one-dimensional case of propagation of a transient
field in a homogeneous infinite medium with conductivity σ, relative
permittivity ϵ and permeability μ.

Wait's general method is to determine the form of the magnetic field and
electric field at a distance from the source point. The transient fields are
represented as Fourier Integrals and in the case where the driving function is a
unit impulse the time domain characteristics of the far field response are derived.
Wait further develops the general approach for the case of a dipole on a
dielectric half space and concludes that the main feature of propagation in
conducting media is that waveform changes its shape as it propagates away from
the source. As a consequence, resolution is severely degraded. However, this loss
of fidelity could also be used as an indicator of distance travelled.

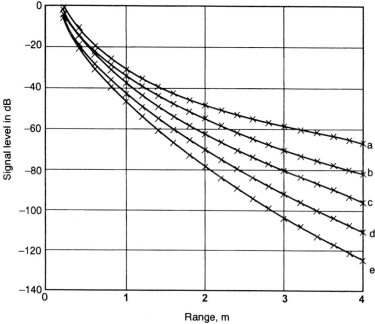

Figure 3.6   Received signal level against range at 1 GHz
    a Attenuation = 3.7 dB/m, loss tangent = 0.02
    b Attenuation = 7.3 dB/m, loss tangent = 0.04
    c Attenuation = 10.9 dB/m, loss tangent = 0.06
    d Attenuation = 14.5 dB/m, loss tangent = 0.08
    e Attenuation = 18 dB/m, loss tangent = 0.1

King and Wu (1993) treat the situation of the propagation of a radar pulse in sea water in similar general manner and draw a parallel to the treatment by Brillouin (1914) and Sommerfeld (1914) of the propagation of optical pulses in a linear, causally dispersive medium and in an updated form by Oughstrun (1991). King considers analytically the case of the near field generated by the rectangular-pulse-modulated current in an electric dipole in sea water. This situation is a useful model for pulse propagation in a dissipative and dispersive media from a dipole source. King concludes that the amplitude of the wave packet decays more rapidly than the amplitudes of the transient components. Finkelshteyn and Kraynyukov (1986) also consider the effect of the medium on pulse propagation characteristics. Wait and Nabulsi (1992) also consider the possibility of performing the pulse shape to suit particular lossy media.

3.3 Properties of lossy dielectric materials

The electromagnetic behaviour of natural and man-made materials is generally complicated because all exhibit both dielectric and conducting properties. Their electromagnetic characteristics are controlled by the microscopic scale (atomic,

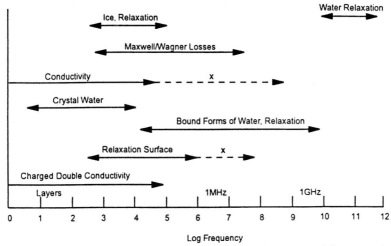

Figure 3.7   Origin of dielectric losses in heterogeneous mixtures containing water (after De Loor, 1983)

molecular and granular) behaviour of the components making up the materials. The origins of various dielectric losses as a function of frequencies are shown schematically in Figure 3.7. The effects occur at different frequencies which create a frequency dependency in the dielectric properties of these materials. Figure 3.7 illustrates the changes which take place in the relative permittivity ϵ_r' and the dielectric-loss factor ϵ_r'', over an extremely wide frequency range. In this idealised representation, the relative permittivity effectively remains constant at high and low frequencies. However, there is a transition region over an intermediate frequency band where the dielectric properties change significantly with frequency. This region is of particular interest when it occurs in the microwave band. In Figure 3.8, the regions of electronic resonance (circa 10^{15}

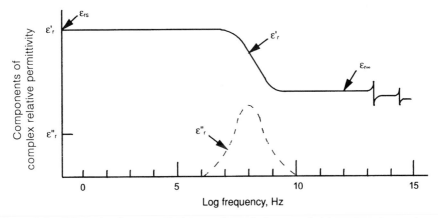

Figure 3.8   Schematic diagram of the complex relative permittivity relaxation region (atomic and electronic resonance regions) $\epsilon_r = \epsilon_r' - j\epsilon_r''$

Hz) occur at frequencies far higher than those associated with surface-penetrating radar. Therefore, they need not be considered any further in relation to the frequency range of interest.

The relaxation phenomenon portrayed relates to the disturbance of polar molecules by an impressed electric field, each molecule experiencing a force that acts to orientate the permanent dipole moment characteristic of the molecule parallel to the direction of the applied electric field. This force is opposed by thermodynamic forces. If an alternating electric field is applied, the individual molecules will be induced to rotate in an oscillatory manner about an axis through their centres, the inertia of the molecules preventing them from responding instantaneously. Similar translational effects can occur. The polarisation produced by an applied field (such as a propagating radar wave) is closely related to the thermal mobility of the molecules and is, therefore, strongly temperature dependent. In general, the relaxation time (which may be expressed as a relaxation frequency) depends on activation energy, on the natural frequency of oscillation of the polarised particles, and on temperature. Relaxation frequencies vary widely between different materials. For example, maximum absorption occurs at very low frequencies in ice (10^3 Hz), whereas it takes place in the microwave region in water (10^6–10^{10} Hz). Thus the effects of this phenomenon could have a direct bearing upon the dielectric properties of materials at the frequencies employed by surface-penetrating radars, especially if moisture is present within a material. There are a number of other mechanisms which cause a separation of positively and negatively charged ions resulting in electric polarisation. These mechanisms can be associated with ionic atmospheres surrounding colloidal particles (particularly clay minerals), absorbed water and pore effects, as well as interfacial phenomenon between particles. This behaviour is illustrated in Figure 3.9 which shows the complex behaviour of a silty clay soil over the frequency range 10^3–10^{11} Hz.

Although the above is a basic description of a complex subject, it does serve to explain the frequency-dependent nature of the dielectric properties of the materials involved. This implies that there will be some variation in the velocity of propagation with frequency. Dielectrics exhibiting this phenomenon are termed dispersive. In this situation, the different frequency components within a broadband radar pulse would travel at slightly different speeds, causing the pulse shape to change with time. However, the propagation characteristics of octave-band radar signals remain largely unaffected and most commercial surface-penetrating-radar systems fall into this category.

The determination of the dielectric properties of earth materials remains largely experimental. Rocks, soils and concrete are complex materials composed of many different minerals in widely varying proportions and their dielectric parameters may differ greatly even within materials which are nominally similar. Most earth materials contain moisture, usually with some measure of salinity. Since the relative permittivity ϵ_r of water is of the order of 80, even small amounts of moisture cause a significant increase of the relative permittivity of the material. An indication of the effect of moisture content on the relative

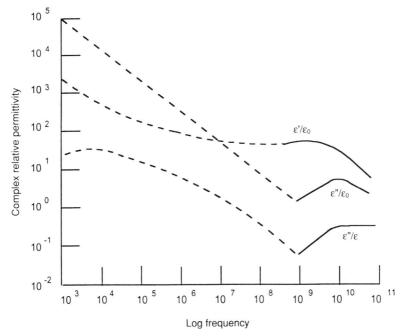

Figure 3.9   The complex relative permittivity and loss tangent of a silty clay soil at a water content of 15% wt (Hoekstra and Delaney, 1974)

permittivity of rock is shown in Figure 3.10. A large number of workers have investigated the relationships between the physical, chemical and mechanical properties of materials and their electrical and, in particular, microwave properties. In general, they have sought to develop suitable models to link the properties of the material to its electromagnetic parameters. Such models provide a basis for understanding the behaviour of electromagnetic waves within these media.

The influence of moisture content on the dielectric properties of earth materials is significant and is well documented in the literature. Extensive measurements at 450 MHz and 35 GHz were made by Campbell and Ulrich (1969) on dry mineral and rock samples. The difference between the apparent permittivity at the two frequencies was small, supporting their conclusion that dry materials have no measurable dispersion at microwave frequencies. The relative permittivity varied from 2.5 for low-density rock types to 9.5 for high-density basaltic rocks, and the loss tangent $(\tan \delta)$ was less than 0.1.

Von Hippel (1954) reports a series of measurements on soils at various water contents up to frequencies of 10 GHz. When the water content of his samples exceeds 10% by weight, a substantial decrease in relative permittivity and an increase in the loss tangent was observed at frequencies between 100 MHz and 10 GHz. This dispersion is considered to be due to the dielectric relaxation of water in soils.

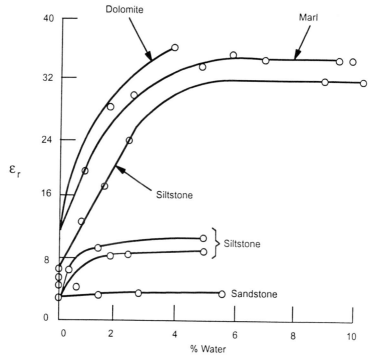

Figure 3.10   Effect of moisture content of rock on relative permittivity (after Hipp, 1974)

There are a number of methods of classifying soils and the Soil Survey of England and Wales classifies soil particles as

Soil textual class	*Size*
Clay	Less than 0.002 mm diameter
Silt	0.002–0.06 mm diameter
Fine sand	0.02–0.06 mm diameter
Médium sand	0.06–0.2 mm diameter
Coarse sand	0.06–2.0 mm diameter
Stones	above 2.0 mm diameter

The description of a soil is arrived at by calculating the relative percentage of silt, clay and sand and the resultant description can be obtained using the triangular diagram shown in Figure 3.11.

Table 3.2 provides an indication of the electromagnetic characteristics of a number of earth materials for typical operating frequencies utilised for surface-penetrating-radar work.

In many instances the potential variation in the velocity of wave propagation over the frequency range of interest would be small and will be ignored. It would only be considered in special circumstances, perhaps where relative indications were required about major variations in the moisture content of a material. In general, it is not possible to make a reliable estimate of propagation velocity or

Table 3.2   Attenuation and relative permittivity of various materials measured at 100 MHz (reported)

Material	Attenuation	Relative permittivity ϵ_r
	dB/m	F/m
Air	0	1
Asphalt: dry	2–15	2–4
Asphalt: wet	2–20	6–12
Clay	10–100	2–40
Coal (UK): dry	1–10	3.5–9
Coal (UK): wet	2–20	8–25
Concrete: dry	2–12	4–10
Concrete: wet	10–25	10–20
Freshwater	0.1	80
Freshwater ice	0.1–2	4
Granite: dry	0.5–3	5
Granite: wet	2–5	7
Limestone: dry	0.5–10	7
Limestone: wet	10–25	8
Permafrost	0.1–5	4–8
Rock salt: dry	0.01–1	4–7
Sand: dry	0.01–1	4–6
Sand: saturated	0.03–0.3	10–30
Sandstone: dry	2–10	2–3
Sandstone: wet	10–20	5–10
Seawater	1000	81
Seawater ice	10–30	4–8
Shale: saturated	10–100	6–9
Snow: firm	0.1–2	8–12
Soil: sandy dry	0.1–2	4–6
Soil: sandy wet	1–5	15–30
Soil: loamy dry	0.5–3	4–6
Soil: loamy wet	1–6	10–20
Soil: clayey dry	0.3–3	4–6
Soil: clayey wet	5–30	10–15

relative permittivity in a medium from a single measurement without trial holing or other supplementary information. Even where a measurement is carried out at one location, it is often found that significant variations in velocity will occur within comparatively short distances from the original location. This can lead to significant errors in the estimation of depths of reflectors or thicknesses of layers. One procedure which overcomes this limitation is known as common-depth-point surveying which utilises two antennas in bistatic operation at a number of transmit and receive positions.

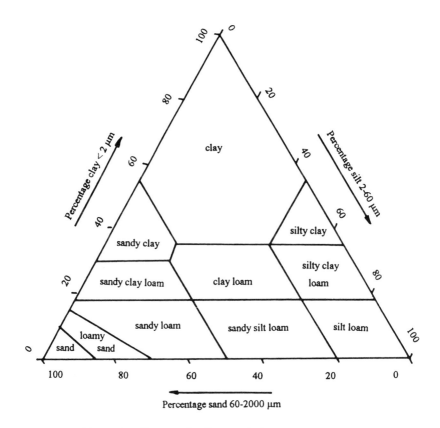

Figure 3.11 Triangular diagram of soil textural classes

In the following Sections the characteristics of the various media that might be investigated by surface-penetrating radar are considered in more detail.

3.4 Water, ice and permafrost

The principal loss mechanism in rocks and soils, certainly at frequencies over 500 MHz, is the absorption of energy by water present in the pores. The dielectric relaxation properties of water are described well by a Debye relaxation with a single relaxation time (Hipp, 1974). Pure liquid water at 0°C exhibits a maximum absorption at 9.0 GHz; this increases as temperature rises, with the maximum absorption occurring at 14.6 GHz at 10°C. Figure 3.12 shows the dielectric relaxation spectrum for water at two temperatures (after Hoekstra and Delaney (1974)).

A useful introductory summary of the properties of water, both liquid and solid, is provided by King and Smith (1981), while Pottel (1973) reviews the topic of

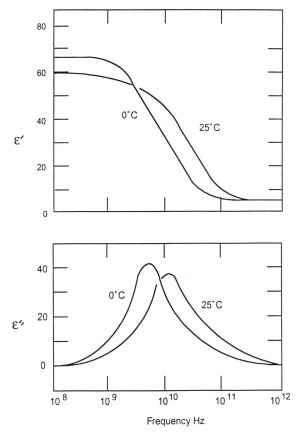

Figure 3.12   Dielectric relaxation spectrum of water at two temperatures (Hoekstra and Delaney, 1974)

aqueous electrolytes. All water encountered in earth materials has some degree of ionic conduction and Pottel's treatment gives useful quantitative information to enable calculations to be made of the frequency dependence of dielectric properties. Cole and Cole (1941), in a classic paper, proposed a modified Debye relaxation equation for liquid water containing free ions. An indication of the variation in permittivity for a solution of NaCl and pure water is for values of ϵ_s, the high frequency permittivity, between 63 and 73 and values of ϵ_∞, the low frequency permittivity, between 5 and 6. It has been found that the quoted range of values for ϵ_∞ is common to water and all aqueous solutions.

The dielectric properties of naturally occurring water (or ice) can be adequately described by the Debye theory or its modification when there is knowledge of its temperature and low-frequency conductivity; with the other constants needed in the calculation being available in the literature. The velocity of wave propagation is slowed substantially with increasing ionic concentrations. This has the most pronounced influence at frequencies below

200 MHz, and above this frequency the wave velocity closely approaches that of pure water. At high conductivities the attenuation rises with frequency and there is no plateau in the loss curve before the rise due to resonance absorption takes place (De Loor, 1983), see Figure 3.8.

An understanding of the dielectric properties of water enables the propagation of electromagnetic waves to be assessed in a range of earth and construction materials. The principal loss mechanism in soils and rocks for frequencies above about 500 MHz is the absorption of energy by water in the pores. If the water content is large enough, the permittivity of the material may be determined first by the dielectric properties of water and to a second order by those of the dry material.

For many applications the real effective permittivity and conductivity may be assumed to be

$$\epsilon'_e = \epsilon' \tag{3.36}$$

$$\sigma'_e = \sigma_{DC} + \omega\epsilon'' \tag{3.37}$$

where σ_{DC} is the DC conductivity ϵ' *and* ϵ'' are obtained from the Debye formula given by King and Smith (1982):

$$\epsilon(\omega) = \epsilon' - j\epsilon'' = \epsilon_\infty + \frac{\epsilon_s - \epsilon_\infty}{1 + j\omega\tau} \tag{3.38}$$

where

$$\epsilon_\infty = \text{high-frequency (infinite) permittivity}$$
$$\epsilon_s = \text{low-frequency (static) permittivity}$$
$$\tau = \text{relaxation time of water}$$

The variation of ϵ' and ε'' for water is presented as a function of frequency in Figure 3.12.

The above assumes a very low pure-water DC conductivity, with an associated low attenuation in the frequencies employed for surface-penetrating radar. In practice, impurities or the presence of free ions in water significantly increase conductivities in the resulting electrolyte. Cole and Cole (1941) proposed a modified Debye relaxation equation for complex permittivity at angular frequency ω:

$$\epsilon(\omega) = \epsilon_\infty + \frac{\epsilon_s - \epsilon_\infty}{1 + (j)^{1-a}(\omega\tau)^{1-b}} \tag{3.39}$$

which reduces to the Debye relation when $a = b = 0$.

An approximately linear relationship was found between the concentration of the ionic solution and the deviation of the relaxation time τ from that of pure water, for up to 1M concentration. It is also found that the DC conductivity σ_{DC}, also varies linearly with concentration in weak solutions.

The complex permittivity of a liquid can be estimated for a given concentration, temperature and frequency on the basis of the equation for $\epsilon(\omega)$ and data obtained from the literature.

Ice is a material which has been extensively measured using surface-penetrating radar, in both the Antarctic and the Arctic. Fresh-water ice displays very different properties from sea-water ice and the latter is considered by Kovacs *et al.* (1987) to be a complex, lossy, anisotropic dielectric consisting of pure ice, air, brine and possibly solid salts.

Generally, sea ice is classified by age into first, second and subsequent years and by structure. The latter is defined by the processes of growth, melt and deformation. Seawater is an electrolyte comprising a multitude of salts of which sodium chloride is the dominant constituent (approximately 80%). The formation of sea ice occurs in several separate phases which govern the geometric structure and hence the anisotropic electromagnetic characteristic of the ice. The initial growth phase of a few millimetres generates needles of ice and these are termed frazil crystals. The second phase of growth occurs when the ice thickens and the initially random crystal orientation becomes more ordered and wider as the depth of ice increases. The ice crystals become vertically orientated as they grow downwards. As growth continues, the elongated crystals cause local freezing of the sea water, effectively localising the water from the base and thus any remaining water in the spaces between the crystals increases its salt content. As the crystals grow, they entrap local pockets of high salinity water. Underwater currents affect the orientation of the crystals and hence the anisotropy of the ice layers which form over many years.

Kovacs *et al.* (1987) calculated that the apparent relative permittivity measured at 100 MHz decreased with increasing ice thickness and followed the trend established with field measurements. Their model of sea ice suggests that the relative permittivity at a frequency of 100 MHz increases in a nonlinear fashion from approximately 3.5 to 18 as the ice depth increases from 0.2 m to 0.7 m. Over the same depth range, the attenuation at 100 MHz increases from approximately 10 dB/m to 50 dB/m.

Kovacs *et al.* (1987) also showed that the coefficient of anisotropy, defined as the ratio of the major to minor axis of the polar plot of the reflection amplitude, could vary between 2.2 and 15. However, the situation he encountered in Alaska was further complicated by the inclusion of different strata within the ice which affected the orientation and growth pattern of the C axis of the ice crystals.

Radar survey work has also been carried out on the detection of ice in permafrost near the Alyeska pipeline in Alaska.

3.5 Dielectric properties of soils and rocks

The dielectric properties of soils and rocks are discussed in detail by De Loor (1983), Hoekstra and Delaney (1974), Hipp (1974), Wang and Schmugge (1980), Hallikainen *et al.* (1985), Wobschall (1977), and extensive data is provided by Parkhomenko (1967), Keller (1971), Fuller and Ward (1977), Campbell and Ulrichs (1969) and Endres and Knight (1992).

All of these authors consider in detail the theoretical effect of microscopic fluid distribution on the dielectric properties of partially saturated rocks. The electromagnetic characteristics of rocks include: anisotropy, an enormous range (20 orders of magnitude) of DC conductivity and an order-of-magnitude range in permittivity ϵ_r between about 2 or 3 for dry sandstone and 40 or so for wet porous rocks. The use of radar techniques to probe rock is usually intended to provide information about deep features where resolution is not usually critical. This permits the use of lower frequencies to aid penetration. Typical changes in relative permittivity with moisture content are shown in Figure 3.10 for several types of rock. These data were obtained at frequencies in the range of 0.1–1 MHz. The main exception to the requirement for long range is the probing of coal seams where resolution is generally more important.

As values of permittivity and velocity of propagation are strongly influenced by the presence of moisture, there is clearly a relationship between these parameters and porosity of the medium. Work upon rocks has shown the following relationships:

(i) For layered material (Parkhomenko, 1967, p. 314) with the electric field applied parallel to the bedding, for a particular frequency f:

$$\epsilon_r = (1 - p)\epsilon_m + p\epsilon_w \tag{3.40}$$

where

$$\epsilon_r = \text{permittivity of layered material}$$
$$\epsilon_m = \text{permittivity of matrix}$$
$$\epsilon_w = \text{permittivity of water}$$
$$p = \text{porosity ratio}$$

(ii) For layer material (Parkhomenko, 1967) with electric field applied perpendicular to the bedding, for a particular frequency, f:

$$\epsilon_r = \frac{\epsilon_m \epsilon_w}{(1 - p)\epsilon_m + p\epsilon_w} \tag{3.41}$$

from which velocity can be calculated. Alternatively, porosity can be calculated from:

$$p = \frac{c}{v^2} = \frac{\epsilon_m}{(\epsilon_m - \epsilon_w)} \tag{3.42}$$

Endres and Knight (1992) have put forward a theoretical treatment of the effect of microscope fluid distribution on the dielectric properties of partially saturated rocks.

Attempts have also been made by Hanai (1969), amongst others, to model the dielectric behaviour of soils by extending the theory of dielectric mixtures produced by Wagner and others for emulsions for the oil-in-water type, establishing relationships of the following type:

$$\frac{(\epsilon^* - \epsilon_p^*)}{(\epsilon_m^* - \epsilon_D^*)} \left(\frac{\epsilon_m^*}{\epsilon^*}\right)^{0.33} = 1 - \phi \tag{3.43}$$

where

$$\phi = \text{volume fraction of the dispersed phase}$$
$$\epsilon_D^* = \text{permittivity of dispersed phase}$$
$$\epsilon_m^* = \text{permittivity of dispersing medium}$$
$$\epsilon^* = \text{permittivity of mixture.}$$

The dielectric properties of soils have been studied for many years and there is now a large body of experimental data available as well as a range of theoretical models. Considerable difficulties are posed by the variability of the material and none of the models developed are universally applicable. Simple models for dielectric loss tend to be deficient, with the major discrepancy between theory and experiment being the frequency dependence of the observable effects. The principal errors are understood to relate to the representation of the energy absorption by moisture, although there are numerous other factors which have a bearing upon the matter.

Experimental studies (Hipp, 1974) upon soils have shown a rise in permittivity with water content and, at a given water content, a fall in permittivity with increasing frequency. Effective conductivity, and hence attenuation, rose both with frequency and water content. Figure 3.9 illustrates the dielectric behaviour of a silty clay soil at a moisture content of 15% by weight over a wide range of frequencies after Hoekstra and Delaney (1974).

Field measurements have produced a wide scatter of results primarily owing to the inherent variability of the 'natural' environment caused by the presence of stones, boulders and localised regions of high conductivity within the 'ground' mass. Such variations cause the dielectric parameters to change in a statistically unpredictable way as the radar antenna is scanned over the 'ground' surface introducing clutter into the received signal. There is no simple parameter, such as water content or low-frequency conductivity, which can be used as a convenient measure of dielectric loss in the frequency range 100 MHz to 1 GHz.

Various workers have sought to employ a modified Debye model for describing the dielectric properties of moist soils. Bhagat and Kadaba (1977) indicate that a relaxation mechanism alone is adequate at frequencies above about 1 GHz, but at lower frequencies the soil structure affects the results. De

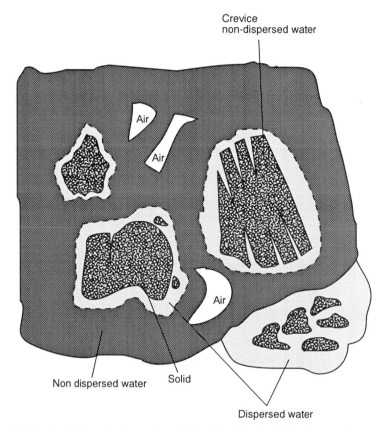

Figure 3.13   Particles and water dispersed in soil (after Wobschall, 1977)

Loor (1983) summarises the chief loss mechanisms occurring in soil and their frequency range of applicability. In comparing the relaxation of water in bulk and soil water, it is observed that the frequency at maximum dielectric loss is displaced to a much lower frequency in soils and that relaxation occurs over a narrower frequency band in soils than in bulk water.

Soil is not only a mixture of dielectrics, but even when the composition of a given sample is known in terms of its components and their individual properties, that alone is not sufficient to define its nature dielectrically. The particle sizes, the electrochemical nature of their boundaries and the way in which the water is distributed, both physically and chemically, amongst the matrix also affect the behaviour. Wobschall (1977) distinguishes, for example, between free water and pore water.

Systematic attempts have been made to produce models of the dielectric behaviour of soils which use parameters obtained independently of the dielectric measurements to be predicted. A number of models are believed to show reasonable agreement with experimental values over certain frequency ranges, but scope remains for further work on the subject of soil dielectric properties.

Wobschall (1977) developed a theory of the complex dielectric permittivity of soils based on the semidisperse model.

This is an extension of a three-phase (particles, air, water) system developed from work carried out by Hanai, Bruggelman and Wagner. Wobschall considers soil to consist of irregular particles containing micro/macroscopic air-filled void (pore) and crevices which become increasingly filled by water as the percentage water content increases. Figure 3.13 shows the schematic composition of the soil suggested by Wobschall. The key element of the semidisperse model is a phase of composition in which the solid particles contain disperse water filled with micropores and around each particle is a coating of water. These particles and the entrapped air are dispersed throughout the remaining water. The procedure for calculation takes into account 10 parameters including the frequency, the volume fraction of water, the volume fraction of voids, the conductivity of pore water and microcrevice water, the dispersed-water fraction and the crevice-water fraction.

There is, however, a sufficient body of information to enable an assessment to be made of the likely range of dielectric properties which might be encountered, assuming there is a detailed knowledge of the site in question.

3.6 Dielectric properties of man-made materials

As the evaluation of concrete structures is of particular importance, more work is now being carried out on establishing the electrical properties of cements and concretes, especially at the higher frequencies associated with surface-penetrating-radar applications. The earliest work which has been carried out has largely been related to DC or low-frequency AC resistivity measurements and the development of models describing the mechanisms of conduction through Portland-cement pastes. Such techniques have been used to study changes in the electrical characteristics of cement pastes and concretes during the stages of initial setting and the subsequent hardening process.

Concrete is a composite material formed by the addition of water to a mixture of cement, sand and coarse aggregate, together with a small quantity of air. A range of admixtures is available which may be added to the mix to modify the properties of concrete in either the fresh or hardened states, or both. A number of different types of cement are employed, although something in the order of 90% of all concrete is produced using ordinary Portland cements. A diverse range of rock types is used to provide the fine and coarse aggregates.

Hydration takes place between the cement and the water in the mix, producing a matrix of hydrates of various compounds, referred to collectively as cement paste, which binds the constituents together to form hardened concrete. The cement paste contains a large number of interconnected voids, the capillary pores, which are the remnants of the water-filled spaces present in the fresh paste. Capillary pores are estimated to be of the order of 1 μm and are mainly responsible for the permeability of the hardened cement paste. The permeability of concrete, and hence the volume of capillary pores, increases rapidly for water : cement ratios over 0.4. The permeability of a mature cement paste made with a water : cement ratio of 0.7 is of the order of 100 times that for one made with a water : cement ratio of 0.4.

After setting, a large amount of the free water within the mix is absorbed during the process of hydration, being used in the production of hydrates of various compounds. Water is also lost from the capillary pores as the concrete dries out. This occurs by evaporation at a rate which is dependent on the water: cement ratio of the concrete, its age, the curing regime to which the member has been subjected and a number of other factors including member size and environmental conditions.

Whittington *et al.* (1981) discuss the conduction of electricity through the heterogenous medium of concrete and report low-frequency AC experimental determinations of changes in resistivity over a three-month period from the time of casting. Their results show resistivity values increasing with time, achieving stable values after about 28 days, although it should be noted that their work was undertaken on moist specimens (i.e. specimens which were not permitted to dry out). It was found that the experimental curves for cement pastes and concretes follow the same trends, confirming the dominant role of the cement paste upon the electrical characteristics of concrete.

The electrical resistivities of typical aggregates particles used in concretes are several orders of magnitude higher than that of the concrete. Consequently, the majority of current flow takes place through the paste (i.e. the path of least resistance). In a simplified electrical model, concrete can generally be considered as a composite of nonconducting particles contained in a conductive cement-paste matrix.

Nikhannen (1962) has suggested that conduction through moist concrete is essentially electrolytic in nature, and tests by Hammond and Robson (1975a and b) and Monfore (1968) support the view that conduction is by means of ions in the evaporable water in the capillary pores, the principal ions being calcium (Ca^{++}), sodium (Na^{+}), potassium (K^{+}), hydroxyl (OH^{-}) and sulphate (SO_4^{-}). Since the amount of evaporable water in a typical cement paste varies from about 60% at the time of mixing to 20% after full hydration, the electrical conductivity of the concrete should also be a function of time. Ionic conduction through the free evaporable water within the capillary pores will depend upon the species, concentration and temperature of the ions present in solution.

Another possible path for current flow is by means of electronic conduction through the cement compounds themselves, i.e. the gel, gel-water and unreacted cement particles, particularly compounds of iron, aluminium and calcium. Wyllie and Gregory (1958) suggest that even physically immobile water will be involved in the conduction process.

Thus, the electrical conductivity of the paste depends upon the changes which both solid and solution phase (i.e. evaporable water) undergo. These changes are closely linked to each other. The composition and concentration of ions in the evaporable water depend upon the soluble compounds within cement particles and the residual water. Conversely, the composition and structure of the solid phase depends upon the amount of water, both absorbed and chemically combined, within the cement compounds during the hydration process.

The above findings for DC and low-frequency AC resistivity measurements indicate the changing nature of the general electrical characteristics of cement pastes and concretes during the first few weeks and months after casting. Further work needs to be carried out to establish the dielectric properties in this period at the higher frequencies associated with surface penetrating radar applications. It is clear that attenuation loss and velocity of wave propagation in concrete are dependent upon a number of factors including water content, mix constituents and the curing regime employed. 'Green' concrete may exhibit high values of both relative permittivity ($\epsilon_r = 10$ to 20) and attenuation loss ($\alpha = 20$ to 50 dB/m at 1 GHz). When hydration and curing are effectively complete, at periods of possibly up to about six months, lower values may be measured such that $\epsilon_r = 4$ to 10 and $\alpha = 5$ to 25 dB/m at 1GHz. Thus, the performance of a surface-penetrating-radar system may be affected considerably by the state of the concrete.

Whittington (1986) has investigated the use of the low-frequency electrical characteristics of concrete as a measure of its mechanical properties. He reports a correlation between electrical resistivity and compressive strength over a limited range of resistivity variation. Such a correlation may be of relevance to surface-penetrating-radar applications; however, considerable research work upon the high-frequency characteristics of concrete would be required to establish its validity.

3.7 Summary

The most important microwave characteristics of soils are the relative permittivity and attenuation. There is an adequate body of literature available on these values as a function of microwave frequency. Further references can be assessed to obtain more information (Al-Attar *et al.*, 1982; Al-Qadi *et al.*, 1991; Annan and Davis, 1978; d'Ambrosio *et al.*, 1991; Evjen, 1948; Hammond and Robson, 1975*a* and 1975*b*; Hayes, 1979; Hayes, 1982; Keller and Licastro, 1959; Mayham and Bailey, 1975; Monfore, 1968; Scott *et al.*, 1967; Scott and Smith, 1992; Smith and King, 1974; Tran and McPhun, 1981; Whittington, 1986; Wyllie and Gregory, 1958). The values can be used to predict the performance of surface-penetrating radars in typical soils with known moisture contents.

However, the actual performance of radars can vary quite considerably owing to the wide variations often encountered in local material conditions. Both the range and the resolution in depth are affected by the attenuation of the material. An approximate empirical relationship is that the achievable depth resolution lies between 10 and 20% of the probing range.

An indication of the likely range of material properties can be seen from

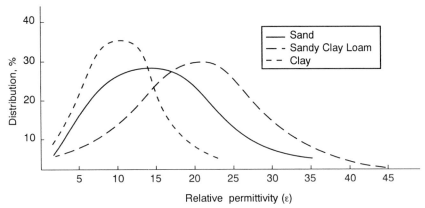

Figure 3.14   *Variation of relative permittivity with soil type over a number of sites in the UK*

Figures 3.14 and 3.15 which refer to the distribution of measurements of relative permittivity and attenuation taken from a range of excavation sites in the UK. Note that for a nominal soil type the relative permittivity can vary considerably. Given a particular frequency, Figure 3.15 shows the percentage of sites with measured attenuations at selected frequencies.

Accurate measurement of depth can only be achieved by proper and frequent calibration of the velocity of propagation. This can be obtained using a multichannel radar system or by secondary measurements. Generally, manmade materials prove to be a more heterogeneous medium for radar probing and silty or heavy clay soils prove difficult to penetrate. Where the material is composed of mixtures containing, for example, rubble, the radar pulses are multiply reflected thus creating a randomly orientated image.

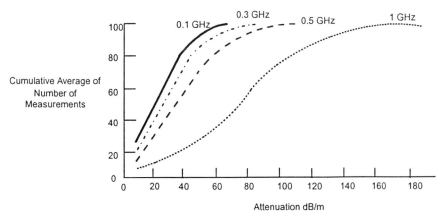

Figure 3.15   *Cumulative variation of attenuation over a number of sites in the UK*

3.8 References

AL-ATTAR, A., SCOTT, H. F., and DANIELS, D. J. (1982): Wideband measurement of microwave characteristics of soils. *Electron. Lett.*, **18**, 194–196

AL-QADI, I. L., GHODGAONKAR, D. K., VARADAN, V. K., and VARADAN, V. V. (1991): Effect of moisture on asphaltic concrete at microwave frequencies. *IEEE Trans. Geosci. Remote Sens.*, **29**, (5), 710–717

ANNAN, A. P., and DAVIS, J. L. (1978): Methodology for radar transillumination experiments. *Curr. Res. B Geol. Surv. Can.*, **78–1B**, 107–110

BHAGAT, P. K., and KADABA, P. K. (1977): Relaxation models for moist soils suitable at microwave frequencies. *Mater. Sci. Eng.*, **28**, 47–51

BRILLOUIN, L. (1914): *Ann. Phys (Leipzig)*, **44**, 177

CAMPBELL, M. J., and ULRICHS, J. (1969): Electrical properties of rocks and their significance for lunar observations. *J. Geophys. Res.*, **74**, 5867–5881

COLE, K. S., and COLE, R. H. (1941): Dispersion and absorption in dielectrics. Alternative current characteristics. *J. Chem. Phys.*, **9**, 341-351

D'AMBROSIO, G., MASSA, R., and MIGLIORE, M. D. (1991): Microwave non destructive testing on composite materials. Proceedings of second international conference on *Electromagnetics in Aerospace Applications*, Torino, Italy, 153-156

DE LOOR, G. P. (1983): The dielectric properties of wet materials. *IEEE Trans.*, **GE–21**, 364–369

ENDRES, A. L., and KNIGHT, R. (1992): A theoretical treatment of the effect of microscopic fluid distribution on the dielectric properties of partially saturated rocks. *Geophys. Prospecting*, **40**, 307–324

EVJEN, H. M. (1948): Theory and practice of low-frequency electromagnetic exploration. *Geophysics*, **13**, 584-594

FINKELSTEIN, M. I., and KRAYNYUKOV, A. V. (1986): Estimation of the delay of super wideband radio pulses in a medium as applied to problems of subsurface radar. English translation of a paper originally published in *Radiotekhnika i elektronika (USSR)*, 1986, (11), 2202–2208

FULLER, B. D., and WARD, S, K. (1970): Linear system description of the electrical parameters of rocks. *IEEE Trans.*, **GE–8**, 7–17

HALLIKAINEN, M. T., ULABY, F. T., DOBSON, M. C., ELRAYES, M. A., and WU, L. K. (1985): Microwave dielectric behaviour of wet soil. Parts 1 and 2. *IEEE Trans.*, **GE–23**, (1), 25–34

HAMMOND, R., and ROBSON, T. D. (1975*a*): Comparison of electrical properties of various cements and concretes. *Engineer (UK)*, **199**, 79–80

HAMMOND, R., and ROBSON, T. D. (1975*b*): Comparison of electrical properties of various cements and concretes. *Engineer (UK)*, **199**, 114–115

HANAI, T. (1969): Electrical properties of emulsions. In SHERMAN, P. (Ed.) (1969): Emulsion science. Academic Press, Chap. 5

HAYES, P. K. (1979): An on-site method for measuring dielectric constant and conductivity of soils over a gigahertz bandwidth. Ohio State University Electroscience Laboratory technical report 582X–1

HAYES, P. K. (1982): A single-probe on-site method of measuring dielectric constant and conductivity of soft earth media over a 1 GHz bandwidth. *IEEE Trans.*, **GE–20**, 504–509

HIPP. J. E. (1974): Soil electromagnetic parameters as functions of frequency, soil density and soil moisture. *Proc. IEEE*, **62**, (1), 98–103

HOEKSTRA, P, and DELANEY, A. (1974): Dielectric properties of soils at UHF and microwave frequencies. *J. Geophys Res.*, **79**, 1699–1708

KELLER, G. V. (1971): Electrical characteristics of the earth's crust *in* WAIT, J. R. (Ed.): *Electromagnetic probing in geophysics*. Golem Press

KELLER, G. V., and LICASTRO, P. H. (1959): Dielectric constant and electrical resistivity of natural-state cores. *US Geol. Surv. Bull.*, **1052-H**, 257–285

KING, R. W. P., and SMITH, G. S. (1981): *Antennas in matter*. MIT Press

KING, R. W. P., and WU, T. T. (1993): The propagation of a radar pulse in sea water. *J. Appl. Phys.*, **73**, (4), 1581–1590

KOVACS, A., *et al.* (1987): Electromagnetic property trends in sea ice. US Department of Energy, CRREL report 87–6

MAYHAN, R. J., and BAILEY, R. E. (1975): An indirect measurement of the effective dielectric constant and loss tangent of typical concrete roadways. *IEEE Trans.*, **AP–23**, 565–569

MONFORE, G. E. (1968): The electrical resistivity of concrete. *J PCA Res. Dev. Lab.*, **10**, (2), 35–48

NIKKANNEN, P. (1962): On the electrical properties of concrete and their application. Valtion Teknillienen Tutkimuslaitos, Tiedotus, Sarja III Rakennus 60 (in Finnish)

OUGHSTRUN, K. E. (1991): *Proc. IEEE*, **79**, 1379

PARKHOMENKO, E. I. (1967): *Electrical properties of rocks.* Plenum Press

POTTEL, R. (1973): Dielectric properties *in* FRANKS, F. (Ed.): Water: a comprehensive treatise. Vol. 3. *Aqueous properties of simple electrolytes.* Plenum Press, Chap. 8

SCOTT, H. H. CARROLL, R. D., and CUNNINGHAM, D. R. (1967): Dielectic constant and electrical conductivity measurements of moist rock: a new laboratory method. *J. Geophys. Res.*, **72**, 5101–5115

SCOTT, W. R., and SMITH, G. S. (1992): Measured electrical constitutive parameters of soil as functions of frequency and moisture content. *IEEE Trans. Geosci. Remote Sens.*, **30**, (3), 621–623

SMITH, G. S., and KING, R. W. P. (1974): The resonant linear antenna as a probe for measuring the in situ electrical properties of geological media. *J. Geophys. Res.*, **79**, (17), 2623–2628

SOMMERFELD, A. (1914): *Ann. Phys. (Leipzig)*, **44**, 203

TRAN, S. M., and McPHUN, M. K. (1981) Obtaining unique solutions to Hanai's equation as used in Wobschall's semidisperse model of moist soil. *IEEE Trans.*, **GE–19**, 69–70

VON HIPPEL, A. R. (Ed.) (1954): *Dielectric materials and applications.* MIT Press, 314

WAIT, J, R. (1960): Propagation of electromagnetic pulses in a homogeneous conducting earth. *Appl. Sci. Res. B. Electrophys. Acoust. Opt.*, **8**, 213–253

WAIT, J, R., and NABULSI, K. A. (1992): Preforming an electromagnetic pulse in lossy medium. *Electron. Lett.*, **28**, (6), 542–543

WANG, J. R., and SCHMUGGE, T. J. (1980): An empirical model for the complex dielectric permittivity of soil as a function of water content. *IEEE Trans.*, **GE–18**, 288–295

WHITTINGTON, H. W. (1986): Low frequency electrical characteristics of fresh concrete. *IEE Proc. A*, **133**, (5), 265–271

WHITTINGTON, H. W., McCARTER, J., and FORDE, M. (1981): The conduction of electricity through concrete. *Mag. Concr. Res.*, **33**, 48-60

WOBSCHALL, D. (1977): A theory of the complex dielectric permittivity of soil containing water. The semidisperse model. *IEEE Trans.*, **GE–15**, 49–58

WYLLIE, M. R., and GREGORY, G. H. F. (1958): The generalised Kozeny-Carmen equation. Part 2. A novel approach to problems of fluid flow. *World Oil*, **146**, (5), 210–228

Chapter 4

Antennas

4.1 Introduction

Surface-penetrating radar presents the system designer with significant restrictions on the types of antennas that can be used. The propagation path consists, in general, of a lossy, inhomogeneous dielectric which, in addition to being occasionally anisotropic, exhibits a frequency-dependent attenuation and hence acts as a lowpass filter. The upper frequency of operation of the system, and hence the antenna, is therefore limited by the properties of the material. The need to obtain a high value of range resolution requires the antenna to exhibit ultrawide bandwidth, and for impulsive-radar systems, linear phase response. The requirement for wide bandwidth and the limitations in upper frequency are mutually conflicting, and hence a design compromise is adopted whereby antennas are designed to operate over some portion of the frequency range 10 MHz–5 GHz depending on the resolution and range specified. The requirement for portability for the operator means that it is normal to use electrically small antennas which consequently result generally in a low gain and associated broad polar-radiation patterns. The classes of antennas which can be used are therefore limited, and the following factors have to be considered in the selection of a suitable design: large fractional bandwidth, low time sidelobes and for separate transmit and receive antennas, low crosscoupling levels. The interaction of the reactive field of the antenna with the dielectric material and its effect on antenna-radiation-pattern characteristics must also be considered.

This Chapter should provide a useful guide to a complex subject but it is not intended to reduce the need for a designer to consult those references cited. In practice, it is found that, provided that an antenna is properly matched, it will couple well into a dielectric. The aspects of coupling, particularly as a function of distance from the interface, are described in detail by the authors cited and for those interested the references should provide a useful source.

Recent interest in ultrawideband radar systems has coincided with the development of additional antenna designs which can provide suitable performance, and much work has been carried out on the development of antennas for ECM and ECCM applications as well as the propagation of high-energy electromagnetic pulses. A useful reference to the general case of antennas in matter is given by King and Smith (1981).

Further considerations in the selection of a suitable type of antenna are the type of target and the type of radar system. Where the target is, for example, a planar surface, linear polarisation is the obvious choice for the system designer. Where, however, the target is a buried pipe or cable, the backscattered field exhibits a polarisation characteristic which is independent of the state of polarisation of the incident field. For linear targets it is possible to use orthogonally disposed transmit and receive antennas as a means of preferential detection. Essentially, the received signal varies sinusoidally with angle between antenna pair and the target. As it is inconvenient to rotate the antenna physically, it is also possible electronically to switch (commutate) the transmit/receive signals to a set of multiple colocated antenna pairs. A further step along this overall strategy is to employ circular polarisation, which is essentially a means of rotating the polarisation vector automatically in space. However, circular polarisation inherently requires an extended time response of the radiated field and in consequence either hardware or software deconvolution of the received signal is needed.

It has been found that very large-diameter pipes exhibit depolarising effects, not from the crown of the pipe but from the edges. The choice of polarisation-dependent schemes should thus be very carefully considered as it may not be possible to cover all possible sizes of target with one antenna/polarisation scheme.

Where the radar system is a time-domain system which applies an impulse to the antenna, the requirement for linear phase response means that only a limited number of types of antenna can be used unless the receiver uses a matched filter to deconvolve the effect of the frequency-dependent radiation characteristics of the antenna. Where the radar system is frequency modulated or synthesised, the requirement for linear phase response from the antenna can be relaxed and log-periodic, horn or spiral antennas can be used as their complex frequency response can be corrected if necessary by system calibration.

Although a full analysis should consider the case of an impulse and consequently a full range of frequencies, it is instructive to examine the case of a single frequency as this provides an understanding as to the resultant radiation patterns of an antenna situated over a halfspace.

The general situation of an antenna adjacent to a lossy halfspace, or indeed fully immersed into lossy materials, has been considered by a number of workers. For an electrically small linear antenna, as shown in Figure 4.1 with a uniform current distribution, the electric- and magnetic-field components in free space are given by

$$H_\phi = \frac{I_o h}{4\pi} e^{-jkr} \left(\frac{jk}{r} + \frac{1}{r^2} \right) \sin \theta \tag{4.1}$$

$$E_r = \frac{I_o h}{4\pi} e^{-jkr} \left(\frac{2\eta}{r^2} + \frac{2}{j\omega e r^3} \right) \cos \theta \tag{4.2}$$

$$E_\theta = \frac{I_o h}{4\pi} e^{-jkr} \left(\frac{j\omega\mu}{r} + \frac{1}{j\omega e r^3} + \frac{\eta}{r^2} \right) \sin \theta \tag{4.3}$$

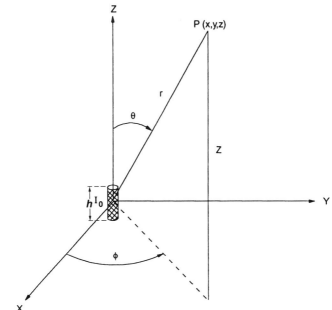

Figure 4.1 Electrically small antenna

In the reactive field of the antenna when r is small the most important term is H_ϕ which varies as $1/r^2$ and in E_r and E_θ those terms which vary as $1/r^3$. At large distances from the source, the terms of E and H which are most significant are those varying as $1/r$. For a halfwave dipole, the far-field intensity

$$|E_\theta| = \frac{60 I_m}{r} \left(\frac{\cos\left(\frac{\pi}{2}\cos\theta\right)}{\sin\theta} \right) \qquad (4.4)$$

The typical radiation pattern is shown in Figure 4.2.

For an aperture antenna, as might be used in an FMCW or stepped-frequency radar, the field at a point $P(x, y, z)$ some distance from the aperture is given by the scalar field $F(P)$ which is known as the Fresnel–Kirchoff scalar-diffraction field.

It is derived by integrating over the aperture area all elements $a\alpha$, $a\beta$ composing the radiating structure

$$F(P) = \frac{1}{4\pi} \int_{area} G(\alpha, \beta) \frac{e^{-jkr}}{r} \left(\left(jk + \frac{1}{r}\right) \cos(n, r) + jk \cos(n, s) \right) d\alpha d\beta \qquad (4.5)$$

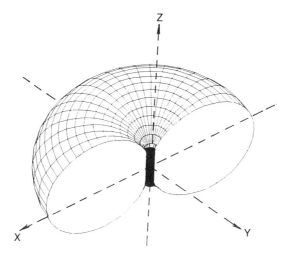

Figure 4.2   Radiation pattern of an electrically small antenna

where $G(\alpha, \beta)$ is the illumination of the field across the aperture

$$k \quad = \quad 2\pi/\lambda$$
$n, r \quad =$ angle between the normal to aperture and the r direction
$n, s \quad =$ angle between the normal to the aperture and the phase
illumination across the aperture.

The expression for $F(P)$ is traditionally separated into three regions depending on the assumptions made to solve the problem.

The first region is defined as the reactive near field and exists close to the aperture. It is generally considered that the above integral formulation is not rigorous as the boundary conditions are undefined.

The Fresnel region is considered to extend from the reactive near field and takes in a radiating near-field region which extends to $2D^2/\lambda$ where D is the aperture's maximum dimension.

Skolnik (1970) details a number of approximations which enable the scalar field in the Fresnel region to be defined as

$$F(P) = \frac{(1 + \cos\theta)}{2\lambda} \frac{e^{-jkz}}{R} \int_{area} G(\alpha, \beta) \exp\left(\frac{(x - \alpha)^2 + (y - \beta)^2}{2z}\right) d\alpha d\beta \quad (4.6)$$

However, these approximations restrict the Fresnel approximations to fields at least greater than 8λ from the antenna aperture.

Therefore any calculations relating to antenna radiation should consider the reactive near field and close-in Fresnel regions as being the most likely zones of operation.

It is important to appreciate the effect of the material in close proximity to the antenna. In general this material, which in most cases will be soil or rocks, or indeed ice, can be regarded as a lossy dielectric and by its consequent loading effect can play a significant role in determining the low-frequency performance of the antenna and hence surface-penetrating radar. The behaviour of the antenna is intimately linked with the material and, for borehole radars, the antenna actually radiates within a lossy dielectric, whereas for the surface-penetrating radar working above the surface the antenna will radiate from air into a very small section of air and then into a lossy halfspace formed by the material. The behaviour of antennas both within lossy dielectrics and over lossy dielectrics has been investigated by Junkin and Anderson (1988), Brewitt-Taylor *et al.* (1981), Burke *et al.* (1983) and Rutledge and Muha (1982), and is well reported. The propagation of electromagnetic pulses in a homogeneous conducting earth has been modelled by Wait (1960) and King and Nu (1993), and the dispersion of rectangular source pulses suggests that the time-domain characteristics of the received pulse could be used as an indication of distance.

The interaction between the antenna and the lossy dielectric halfspace is also significant, as this may cause modification of the antenna-radiation characteristics both spatially and temporally and should also be taken into account in the system design. For an antenna placed on an interface, the two most important factors are the current distribution and the radiation pattern. At the interface, currents on the antenna propagate at a velocity which is intermediate between that in free space and that in the dielectric. In general the velocity is retarded by the factor $\sqrt{\{(e_r + 1)/2\}}$.

The net result is that evanescent waves are excited in air whereas in the dielectric the energy is concentrated and preferentially induced by a factor of $\eta^3 : 1$.

The respective calculated far field power-density patterns in both air and dielectric are given by Rutledge and Muha (1982), given in Table 4.1 and plotted for relative permittivities of 4, 6, 8 and 10 in Figures 4.3 and 4.4.

The expressions in Table 4.1 assume that the current source contacts the

Table 4.1 Power-density patterns in air and dielectric

Plane	Power	
	$S(\theta_a)$ Radiation pattern in air	$S(\theta_d)$ Radiation pattern in dielectric
H	$\alpha\left(\dfrac{\cos\theta_a}{\cos\theta_a + \eta(\cos\theta_d)}\right)^2$	$\alpha\eta\left(\dfrac{\eta\cos\theta_d}{\lvert\cos\theta_a + \eta\cos\theta_d\rvert}\right)^2$
E	$\alpha\left(\dfrac{\cos\theta_a\cos\theta_d}{\eta\cos\theta_a + \cos\theta_d}\right)^2$	$\alpha\eta\left(\dfrac{\eta\cos\theta_a\cos\theta_d}{\eta\cos\theta_a + \cos\theta_d}\right)^2$

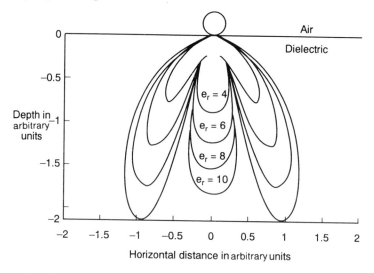

Figure 4.3   E-plane plot of the far-field power density of a current element radiating into a dielectric

dielectric whereas a more general condition is when the antenna is just above the dielectric. The sidelobes in the pattern are a direct result of reactive field coupling. A significant practical problem for many applications is the need to maintain sufficient spacing to avoid mechanical damage to the antenna. It can therefore be appreciated that the effect of changes in distance between the antenna and halfspace cause significant variation in the resultant radiation

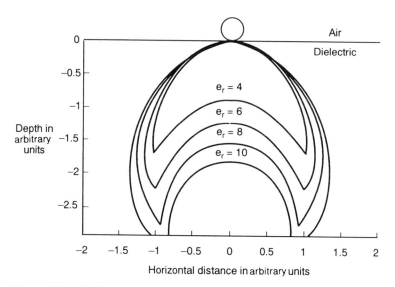

Figure 4.4   H-plane plot of the far-field power density of a current element radiating into a dielectric

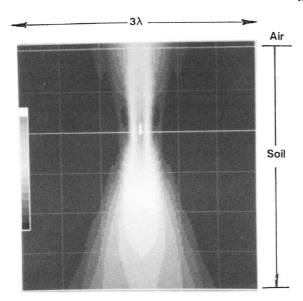

Figure 4.5   Reconstruction of a buried line scatterer for the model scanning at the height of
0.0625λ (courtesy Junkin)

patterns in the dielectric. Where the source–interface space is increased, the antenna field patterns are modified by a reduction in the effect of the reactive field.

This situation was considered by Junkin and Anderson (1988) and the radiation characteristics of the antenna can be considered by examining two conditions: the first is where a line source is placed above and radiates into a halfspace (Figure 4.5) and the other case where a line source is placed within and radiates out of a halfspace (Junkin and Anderson, 1988). For a source above a halfspace where the source–interface spacing is very small, the reactive fields are capable of inducing currents in the halfspace dielectric which subsequently become propagating waves. The situation of the waves reflected from a target can be considered from the point of view of a line source embedded in the dielectric halfspace. This radiated field in free space is shown in Figure 4.6 (Junkin and Anderson, 1988) and it can be seen that surface waves play a significant role.

The overall configuration is further complicated by the use of separate transmit and receive antennas which causes a convolution of the separate radiation patterns to form a composite pattern.

The use of separate transmit and receive antennas is dictated by the difficulty associated with operation with a single antenna which would require an ultrafast transmit–receive switch. As it is not yet possible to obtain commercially available ultrafast transmit–receive switches to operate in the subnanosecond region with sufficiently low levels of isolation between either transmit and receive ports or breakthrough from the control signals, most surface-penetrating-

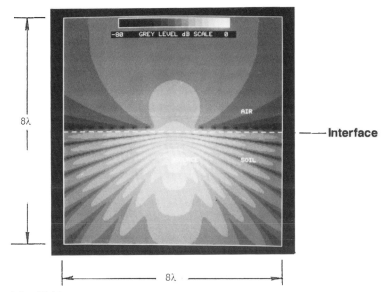

Figure 4.6 *Field pattern of a buried-line source in a material* $e_r = 16\alpha = 3.5$ dB/m, $D = \lambda$ *(courtesy Junkin)*

radar systems use separate antennas for transmission and reception in order to protect the receiver from the high level of transmitted signal. Therefore the crosscoupling level between the transmit and receive antenna is a critical parameter in the design of antennas for surface-penetrating radar and satisfactory levels are usually achieved by empirical design methods. Typically, a parallel-dipole arrangement achieves a mean isolation in the region of -50 dB whereas a crossed-dipole arrangement can reduce levels of crosscoupling to -60 dB to -70 dB. For the crossed-dipole arrangement such levels are highly dependent on the standard of mechanical construction and a high degree of orthogonality is necessary.

The crossed dipole is sensitive to variations in antenna–surface spacing and it is important to maintain the plane of the antenna parallel with the plane of material surface to avoid degrading the isolation.

In the following Sections we shall consider the various types of antenna that can be used with surface-penetrating radar. The types of antenna that are useful to the designer of surface-penetrating radar fall into two groups: dispersive antennas and nondispersive antennas.

Examples of dispersive antennas that have been used in surface-penetrating radar are the exponential spiral, the Archimedean spiral, the logarithmic planar antenna, the Vivaldi antenna and the exponential horn. Examples of nondispersive antennas are the TEM Horn, the bicone, the bow-tie, the resistive, lumped element loaded antenna or the resistive, continuously loaded antenna. A typical antenna used in an impulse-radar system would be required to operate over a frequency range of a minimum of an octave and ideally at least a decade, for example, 100 MHz–1 GHz. The input-voltage driving function to the terminals of the antenna in an impulse radar is typically a Gaussian pulse and this requires the impulse response of the antenna to be extremely short. The main reason for requiring the impulse response to be short is that it is important that the antenna does not distort the input function and generate time sidelobes. These time sidelobes would obscure targets that are close in range to the target of interest, in other words the resolution of the radar can become degraded if the impulse response of the antenna is significantly extended. All of the antennas used to date have a limited low-frequency performance unless compensated and hence act as highpass filters; thus the current input to the antenna terminals is radiated as a differentiated version of the input function.

In general, it is reasonable to consider that the far-field radiated electric field is proportional to the derivative of the antenna current.

It has been shown that, for a physically realisable causal pulse, both the initial value of the antenna current and the initial value of the first derivative of the antenna current must be zero.

If we assume that

$$i(t) = I\left(\frac{t^2}{T}\right)e^{-2t/T} \tag{4.7}$$

then

$$e(t) = k\frac{di(t)}{dt}$$

The current waveform and the radiated electric field are shown in Figure 4.7.

The following sections of this chapter will consider the various classes of antennas that can be used successfully. While these classes can be subdivided into the classes of dispersive antennas and nondispersive antennas, there are actually significant differences in design and in operation of different types within these broad categories. Therefore this Chapter considers element antennas, travelling-wave antennas, frequency-independent antennas, aperture antennas and array antennas separately. It is useful, however, to consider the general requirements of the overall class of antennas which broadly can be considered as being frequency independent. There are a number of requirements for frequency-independent operation and these are as follows:

(i) excitation of the antenna at the region of the antenna from which high frequencies are radiated
(ii) a transmission region formed by the inactive part of the antenna between the feed point and the active region. This zone should produce negligible far-field radiation.
(iii) an active region from which the antenna radiates strongly because of an appropriate combination of current magnitude and phases
(iv) an inactive region created by means of reflection or absorption beyond the active region. It is essential that there is a rapid decay of currents beyond

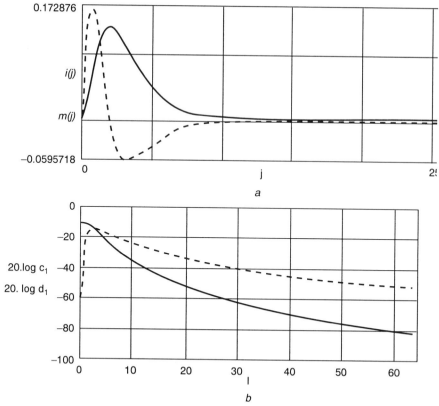

Figure 4.7   Time-domain and frequency-domain response of an antenna when driven by an impulse
——— input
- - - Output
a Time-domain characteristic
b Frequency-domain characteristic

the active region. Efficient antennas achieve this by means of radiation in the active region whereas the less efficient use resistive-loading techniques to achieve this characteristic

(v) a geometry defined entirely by angles, i.e. the biconical dipole, conical spiral, planar spiral, are all defined by angle. These antennas maintain their performance over a frequency range defined by the limiting dimensions. Provided that in these cases, an extended impulse response is acceptable, these antennas provide a performance that can be useful.

4.2 Element antennas

Element antennas such as monopoles, dipoles, conical antennas and bowtie antennas have been widely used for surface-penetrating-radar applications.

Generally, they are characterised by linear polarisation, low directivity and relatively limited bandwidth, unless either end-loading or distributed-loading techniques are employed, in which case bandwidth is increased at the expense of radiation efficiency. Various arrangements of the element antenna have been used such as the parallel dipole and the crossed dipole which is an arrangement that provides high isolation and detection of the crosspolar signal from linear reflectors.

It is useful to consider those characteristics of a simple normally conductive dipole antenna which affect the radiation response to an impulse applied to the antenna feed terminals. As shown in Figure 4.8 two current and charge impulses

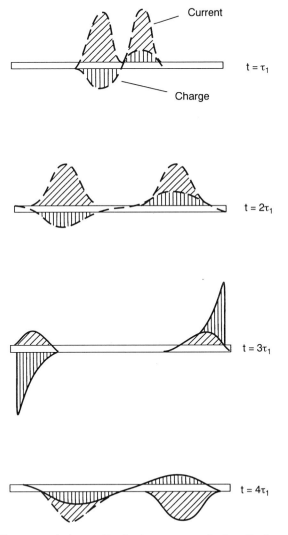

Figure 4.8   Current and charge distribution on a conducting dipole antenna due to an applied impulse

will travel along the antenna elements until they reach each end. At the end of the antenna the charge impulse increases while the current collapses. The charge at the end of the antenna gives rise to a reflected wave carried by a current travelling back to the antenna feed terminals. This process continues for a length of time defined by the ohmic losses within the antenna elements. As far as the radiation field is concerned the relevant parameters (electric-field strength, displacement current and energy flow) can be derived from consideration of Maxwell's equation.

The electric-field component E_z is given by Kappen and Mönich (1987):

$$E_z = -\frac{1}{(4\pi\epsilon_0)} \int_{l_1}^{l_2} \left\{ \frac{1}{c^2}\frac{dl}{dt} + \frac{\partial q}{\partial z'} \right\} \frac{1}{r} dz' \tag{4.8}$$

and must equal zero at the surface of the antenna. This condition can only be satisfied at certain points along the element and implies that, for a lossless antenna, there is no radiation of energy from the impulse along the element. The radiated field is therefore caused by discontinuities, i.e. the feed point and end point are the prime sources of radiation. As would be expected, the time sequence of the radiated field can be visualised by the electric field lines, as shown in Figure 4.9.

Figure 4.9   Radiated field pattern from a conducting dipole element due to an applied impulse

As it is required to radiate only a single impulse, it is important to eliminate either the reflection discontinuities from the far end of the antenna by end loading or reduce the amplitude of the charge and current reaching the far end. The latter can be achieved either by resistively coating the antenna or by constructing the antenna from a material such as Nichrome which has a defined loss per unit area. In this case the antenna radiates in a completely different way as the applied charge becomes spread over the entire element length and hence the centres of radiation are distributed along the length of the antenna.

In essence, the electric field E_z must now satisfy the condition

$$-\frac{1}{(4\pi E_0)} \int_{l_1}^{l_2} \left\{ \frac{1}{c^2} \frac{dl}{at} + \frac{\partial q}{\partial z} \right\} \frac{1}{r} dz' = R'I \tag{4.9}$$

which implies that some dispersion takes place.

The electric-field lines for the lossy element are now different from the lossless case and are shown in Figure 4.10. Further analysis of the radiation characteristics of a resistively coated dipole antenna is given by Randa *et al.* (1991) and Esselle and Stuchly (1990).

The parameters of the antenna such as input resistance, resistivity profile etc. have all been extensively treated in a classic paper by Wu and King (1965). Lumped-element resistors can be placed at a distance $\lambda/4$ from the end of the antenna (Altschuler, 1961) and a travelling-wave distribution of current can be produced by suitable values of resistance. The distribution of current varied

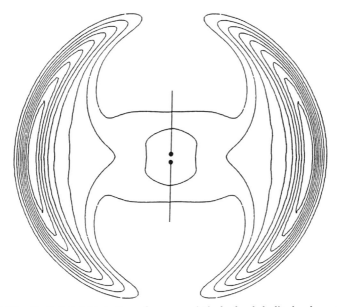

Figure 4.10   *Radiated field pattern from a resistively loaded dipole element due to an applied impulse*

almost exponentially with distance along the element. Instead of lumped-element resistors, a continuously distributed constant internal impedance per unit length can be used. The parameters of a centre-fed cylindrical antenna can be characterised by a distribution of current equivalent to a travelling wave.

The cylindrical antenna with resistive loading has been shown by Wu and King (1965) to have the following properties:

(a) The far-field pattern of the antenna comprised both real and imaginary components, i.e.

$$F_m = \sqrt{\left(F_r^2 + F_i^2\right)} \qquad (4.11)$$

where, for a quarter-wave antenna,

$$F_r = \frac{1 + \cos^2\theta - 2\cos\theta\sin\left(\frac{\pi}{2}\cos\theta\right)}{\left\{\frac{\pi}{2}\right\}\sin^3\theta} \qquad (4.12)$$

$$F_i = \frac{-\frac{\pi}{2}\sin^2\theta + (1 + \cos^2\theta)\cos\left(\frac{\pi}{2}\cos\theta\right)}{\frac{\pi}{2}\sin^3} \qquad (4.13)$$

(b) The efficiency of the antenna is given

$$\eta = \frac{P_r}{P_r + P_a} \qquad (4.14)$$

where

$$P_r = \text{radiated power}$$
$$P_a = \text{absorbed power.}$$

For a resistively loaded antenna of the Wu–King type, the efficiency is approximately 10% but rises to a maximum of 40% for antenna lengths of 40λ.

(c) The resistivity taper profile for a cylindrical monopole has the form given by Rao (1991):

$$R(z) = \frac{R_0}{1 - z/H} \qquad (4.15)$$

where

$$R_0 = \text{resistivity at the drive point of the element}$$
$$H = \text{element length}$$
$$z = \text{distance along the antenna.}$$

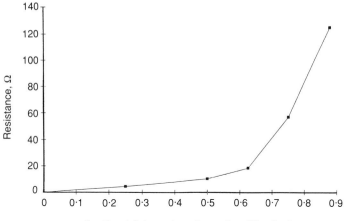

Figure 4.11 *Resistivity profile of a 200 mm element*

A plot of a resistivity against length for a 200 mm element is shown in Figure 4.11. The overall efficiency of this type of antenna can be improved by reducing the value of R_0, and an increase of from 12% to 28% by reduction of R_0 to 0.3 R_0 was shown by Rao. Typical graphs of return loss and crosscoupling are shown in Figures 4.12 and 4.13. The typical time-domain response of a continuously resistively loaded dipole is shown in Figure 4.14.

Further improvement in bandwidth can be gained by matching the antenna with a compensation network and a field probe has been developed with a bandwidth of 20 MHz to 10 GHz by Esselle and Stuchly (1990). Obviously the use of a compensation network further reduces efficiency, but, with a high-impedance receiver probe, a frequency range of 10 MHz–5 GHz can be achieved.

Resistively loaded dipoles have been used as electric-field probes for EMC-measurement applications, and although the frequencies of operation are well in excess of those used for surface-penetrating-radar applications, it is useful to consider the general approach adopted by Maloney and Smith (1991).

Antennas have been developed by Kanda, initially using 8 mm dipoles to measure frequencies up to 18 GHz and subsequently 4 mm dipoles were used by Kanda (Randa *et al.*, 1991) to measure over the frequency range 1 MHz–40 GHz with an error of ±4 dB. The transfer function of this antenna is in the order of −50 dB which illustrates the penalty which is paid for ultrawidebandwidth operation.

A design which offers improved efficiency over the continuously loaded resistive antenna is based on a pair of segmented-blade antennas arranged in a butterfly configuration and fed in phase. Each blade consists of a series of concentric conducting rings connected together by chip resistors. Radial cuts are used to reduce transverse currents. The efficiency of this class of antenna is higher than the continuously loaded dipole without serious degradation of the time-domain response.

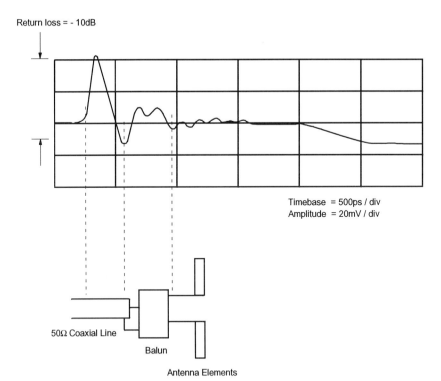

Figure 4.12   *Time-domain-reflectometer return loss of a resistively loaded dipole*
Timebase = 500 ps/division
Amplitude = 20 mV/division

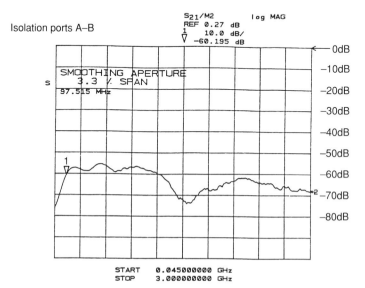

Figure 4.13   *Crosscoupling of a resistively loaded parallel-dipole antenna*

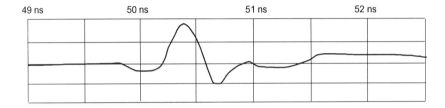

49 ns   50 ns   51 ns   52 ns

Timetrace = 500ps / div
Amplitude = 20mV / div

Figure 4.14 Free-space impulse response of nonresonant target using resistively loaded dipoles
Timebase = 500 ps/division
Amplitude = 20 mV/division

The triangular bowtie antenna has been widely used in commercial surface-penetrating-radar systems. A triangular bowtie dipole of 35 cm length with a 60° flare angle can provide useful performance over an octave bandwidth of 0.5–1 GHz with a return loss of better than 10 dB as shown by Brown and Woodward (1952). Evidently, without some form of end loading, such an antenna would not be immediately suitable for use with impulse-radar systems and the triangular antenna normally uses end loading to reduce the ringing that would usually occur in an unloaded triangular plate antenna. The technique can also be used with a folded dipole and the use of terminating loads results in a transient response equivalent to 1.5 cycles (Young *et al.*, 1977).

An alternative approach to antenna design is based on electric dipole radiation from a Hertzian magnetic dipole as developed by Harmuth (Harmuth, 1983; Harmuth and Ding-Rong, 1983*a* and *b*) provided that the radiation components from current flowing in one direction can be isolated. The advantage of using the Herzian magnetic dipole lies in removing the need, as in the case of the electric dipole, to produce large currents and charges which do not contribute significantly to the far-field radiation field.

In practice, the generator must be shielded, and absorbing ferrite is used to reduce currents flowing on the shield. The main advantage of the Herzian magnetic dipole is the ability to produce very short pulse from electrically small antennas.

4.3 Travelling-wave antennas

In this Section we shall consider the use of antennas capable of supporting a forward-travelling TEM wave. In general, such antennas consist of a pair of conductors either flat, cylindrical or conical in cross-section forming a V structure in which radiation propagates along the axis of the V structure as shown in Figure 4.15. Although resistive termination is used, this type of

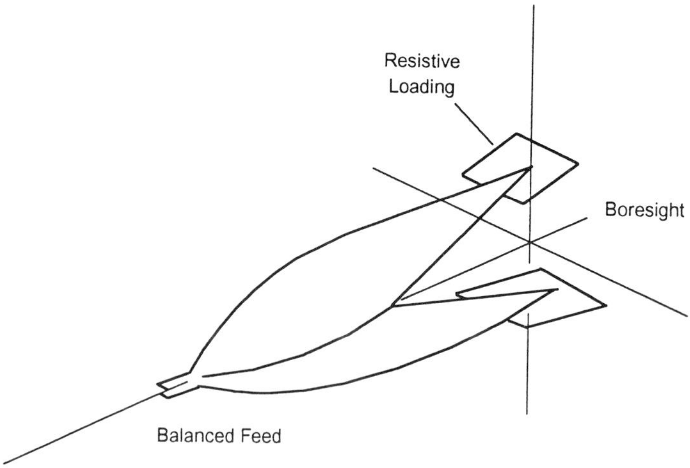

Figure 4.15   Travelling-wave TEM antenna

antenna has a directivity in the order of 10–15 dB; hence useful gain can still be obtained even with a terminating loss in the order of 3–5 dB. The travelling-wave current on one of the cylindrical elements of a V antenna is given by Izuka (1961) and is also discussed by Wu and King (1965):

$$I_t = I_a \, e^{-j\beta z} \tag{4.16}$$

Hence the azimuthal radiation field E is given by

$$E = \frac{j\omega\mu}{4\pi} \frac{e^{-j\beta R}}{R} \int_0^l I_0 \, e^{-j\beta z(1-\cos\theta)} \sin\theta dz \tag{4.17}$$

where

$$
\begin{aligned}
R &= \text{loading resistance} \\
l &= \text{length of the element} \\
z &= \text{distance from radiating source} \\
\theta &= \text{angle in } H \text{ plane}
\end{aligned}
$$

This simplifies to

$$E_T(\theta) = \frac{1 - \exp\{-j\beta(l - l_1)(1 - \cos\theta)\}}{1 - \cos\theta} \sin\theta \tag{4.18}$$

However, a standing wave also exists caused by the resistive termination at the end of the antenna and the contribution from this is given by

$$E_S(\theta) = j \frac{\exp\{-j\beta(l - l_1)\cos\theta\}}{\sin\theta} \left[\cos\left(\frac{\pi}{2}\cos\theta\right) + j\left\{\sin\left(\frac{\pi}{2}\cos\theta\right) - \cos\theta\right\}\right]$$

$$(4.19)$$

The resultant field from one element can be derived from the sum of the contributions

$$E'(\theta) = E_T(\theta) + E_S(\theta) \qquad (4.20)$$

and hence the field from both elements is

$$E = E'_u(\theta) - E_l(\theta) \qquad (4.21)$$

where

> u denotes the upper element
>
> l denotes the lower element.

The antenna will in fact radiate an impulse which is extended in time as a consequence of the geometry of the antenna. The pulse distortion on boresight is given by Theodorou *et al.* (1981):

$$t = \frac{L}{u}(1 - \cos\alpha) \qquad (4.22)$$

where α is the halfangle between the elements and L is the element length and u is the phase velocity of waves along the antenna.

Evidently a small flare angle and short element length help in reducing pulse extension.

The electric field on boresight is related to the time derivative of input current and on boresight is given by

$$E = \frac{-\mu_0 L \sin\alpha}{2\pi r} \frac{\partial I_1\{t - (r/u)\}}{\partial t} \qquad (4.23)$$

The impedance of the antenna should vary in such a way that the derivative of impedance at the feed and end parts is a minimum and along the antenna is low.

Typically, the characteristic impedance is given as a function of distance x as

$$Z_0(x) = Z_x \exp(-K_1 \cos K_2 x) \qquad (4.24)$$

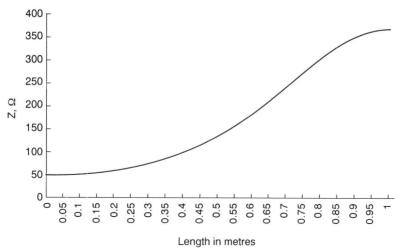

Figure 4.16   Characteristic impedance of a travelling-wave antenna

Hence

$$\frac{d\mathcal{Z}_0(x)}{dx} \to 0 \text{ for } K_2 x = 0 \vee \pi \tag{4.25}$$

Usually the feed impedance is of the order of 50 Ω and the end impedance is desired to be equal to that of free space (377 Ω). However, there is usually a difference between the transmission-line-wave impedance characteristic and that of wave in free space, and a design to meet given criteria in terms of return loss must take this effect into account.

Plots of both typical antenna impedance and rate of change of impedance as a function of length are shown in Figures 4.16 and 4.17.

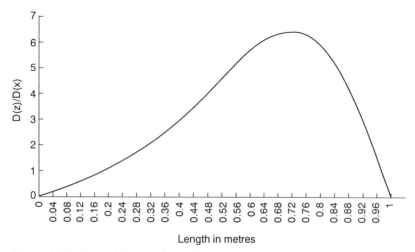

Figure 4.17   Rate of change of impedance of a travelling-wave antenna

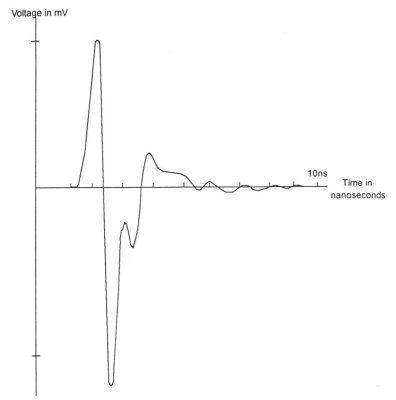

Figure 4.18   *Time-domain response of a pair of travelling-wave antennas used in a face-to-face configuration*

Using this characteristic, a typical antenna–antenna time-domain response is shown in Figure 4.18.

Improved directivity can be obtained using a V-conical antenna as shown by Shen *et al.* (1988). This is formed by a pair of triangular metal plates bent around a cone. The antenna is characterised by two angles, the flare halfangle θ and the azimuthal angle ϕ.

Further developments of the TEM-horn design from the original design first described by Wohlers (1970) are to be found in papers by Daniels (1980), Evans and Kong (1993), Reader *et al* (1985) and Tun (1993).

4.4 Frequency-independent antennas

This class of antennas has a geometry entirely defined by angles and exhibits a performance over a range of frequencies set by the overall dimensions of the structure. Typical examples are the biconical dipole, equiangular spiral and conical spiral. Log-period structures can also provide broadband performance but are not completely defined in terms of angles, Rumsey (1966).

Various developments of the spiral antenna or conical spiral antenna have been carried out by Miller and Landt (1977), Pastol *et al.* (1990), Dyson (1959), Morgan (1985), Kooy (1984), Deschamps (1959), Bawer and Wolfe (1960) and Goldstone (1983).

The impulse response of this class of antennas is extended and generally results in a 'chirp' waveform if the input is an impulse. The main reason for this is that the high frequencies are radiated in time before the low frequencies as a result of the time taken for the currents to travel through the antenna structure and reach a zone in which radiation can take place.

The geometry of the equiangular spiral is defined by

$$\rho = \kappa \, e^{a\phi} \tag{4.26}$$

as shown in Figure 4.19.

The two-arm equi-angular planar spiral can provide acceptable radiation patterns which can be obtained with spirals of as little as 1.25–1.5 turns.

For a planar equiangular spiral, the radiation pattern is bidirectional with equal lobes both front and back of the plane of the antenna. Unidirectional radiation can be achieved by backing the spiral with absorptive material on one side.

The near fields along the arms decay rapidly by as much as 20 dB per wavelength and this reduction is a constant function of the ratio of electrical

Figure 4.19   Equiangular spiral antenna

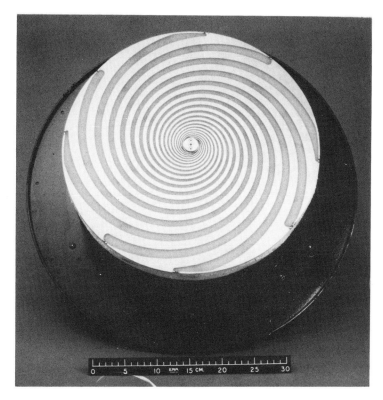

Figure 4.20   Photograph of multi-arm spiral antenna (courtesy ERA Technology)

length of the arm. Effectively, the active arm length is a constant as frequency is increased; hence the effective aperture of the antenna increases with frequency. This characteristic of the equiangular spiral can also be viewed as a nonstationary phase centre which consequently causes dispersion of any impulsive signal applied to the feed terminals.

It is evident that at wavelengths which are of the same order as the length of the arms' the polarisation of the radiated field is linear and as the frequency is gradually increased becomes elliptical and then circular.

A photograph of a multi-arm equiangular spiral is shown in Figure 4.20. In this realisation, eight separate arms are used to form a transmit pair arranged orthogonally to a receive pair and interleaved with screening arms to improve the isolation between the transmit and receive arms. Loading resistors were used to reduce late time currents. The radiation pattern of this antenna at 500 MHz is shown in Figure 4.21 and the on-axis axial ratio is shown in Figure 4.22.

In general, the upper frequency of operation is defined by the accuracy of construction at the feed point or, where the antenna is fed by a balun, the characteristics of the balun.

Where such antennas are excited by an impulsive input waveform, the far-

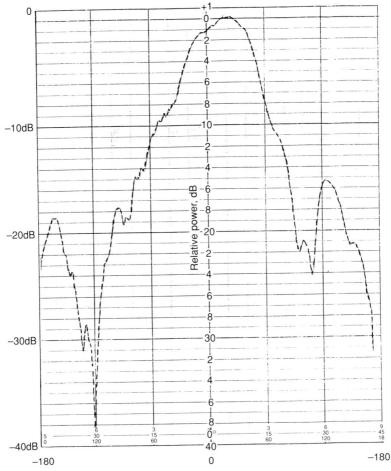

Figure 4.21   Radiation pattern of a single-arm of a multi-arm spiral at 900 MHz

field radiated waveform exhibits significant dispersion. The effect of dispersion can be corrected by deconvolution of the antenna response. A typical time domain characteristic of the equiangular spiral is shown in Figure 4.23.

The short-pulse-radiation characteristics of a conical spiral depend on the type of input waveform, which must be carefully selected to restrict the amount of very low-frequency energy (Goldstone, 1983). If, for example, a Gaussian impulse is applied, then energy at very low frequencies becomes trapped in the antenna which then functions as a resonant structure and radiates an extended far-field waveform. Generally, the antenna must be properly loaded to reduce radiation from such resonant currents.

The main potential advantage of the planar equiangular spiral is the radiation of circular polarisation. Where the target, such as a pipe or cable,

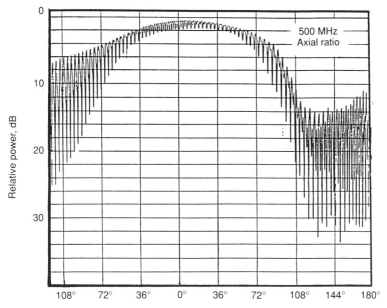

Figure 4.22   Axial ratio at 500 MHz for an equiangular spiral antenna

displays significant polarisation attributes, circular polarisation can be a means of preferential detection.

The Vivaldi antenna (Gibson, 1979) also falls into the class of a periodic continuously scaled antenna structure and within the limiting size of the structure has unlimited instantaneous frequency bandwidth. It provides end-fire

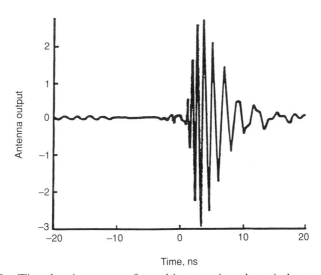

Time, ns

Figure 4.23   Time-domain response of a multi-arm equiangular spiral antenna

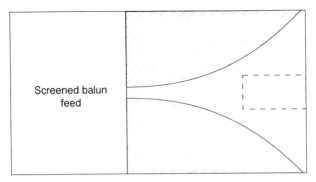

Figure 4.24   Vivaldi antenna

radiation and linear polarisation and can be designed to provide a constant gain frequency of performance. The Vivaldi antenna consists of a diverging slot-form guiding conductor pair as shown in Figure 4.24. The curve of one of the guiding structures follows the equation

$$z = A \; e^{kx} \tag{4.27}$$

Radiation is produced by a nonresonant travelling-wave mechanism by waves travelling down a curved path along the antenna. Where the conductor separation is small, the travelling-wave energy is closely coupled to the conductor but becomes less so as the conductor separation increases. The Vivaldi antenna provides gain when the phase velocity of the travelling wave on the conductors is equal or greater than that in the surrounding medium.

The lower cutoff frequency is defined by the dimensions of the conductor separation, being a half wavelength, and the gain is proportional to overall length.

The impulse response of the antenna is extended due to the nonstationary phase centre but can, of course, be corrected by the use of a matched filter.

4.5 Horn antennas

Horn antennas have found most use with FMCW surface-penetrating radars where the generally higher frequency of operation and relaxation of the requirement for linear phase response permit the consideration of this class of antenna.

Exponentially flared TEM horns with dielectric loading have been developed to operate over decade bandwidths (Kerr, 1973).

The design of horn antennas is well covered in the literature, but of particular interest is the short-axial-length double-ridged horn as shown in Figure 4.25. This design can provide useful gain over a decade bandwidth using a logarithmic characteristic curve for the ridges. Typically the short-axial-length

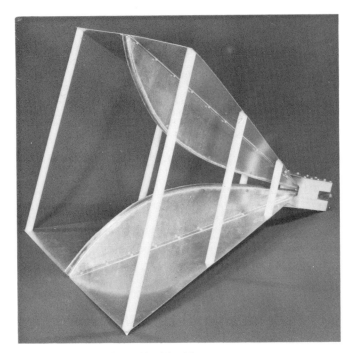

Figure 4.25   Short-axial-length double-ridged horn antenna

horn provides a VSWR of better than 2:1 and a gain of 10 dB over a frequency band of 0.2–2 GHz for an axial length of 0.76 m. Typical radiation patterns are shown in Figures 4.26 and 4.27 The concept of the ridged-horn design can be adapted to form a quad-ridged horn operating from 0.3 to 1.9 GHz. A return loss of better than 10 dB and a crosscoupling level of better than 35 dB can be obtained. The quad-ridged horn can be used to extract information on the polarisation state of the reflected signal.

An FMCW radar has been developed using an offset paraboloid fed by a ridged horn (Sun and Rusch, 1991). The arrangement was designed to focus the radiation into the ground at a slant angle to reduce the level of the reflection from the ground. Care needs to be taken in such arrangements to minimise the effect of back and side lobes from the feed antenna which can easily generate reflections from the ground surface.

Although horn antennas have been mostly used with FMCW systems, it is possible to radiate pulses, and the impulse response of a typical exponential horn antenna is shown in Figure 4.28. Note that pulse shape is distorted by the limited low-frequency performance due to the physical size of the antenna. In addition the exponential flare causes dispersion of the transmitted pulse.

The main advantage of the ridged horn is high aperture efficiency, although at high frequencies where the aperture is many wavelengths wide, large phase errors will be present across the aperture unless the horn is long. The basic

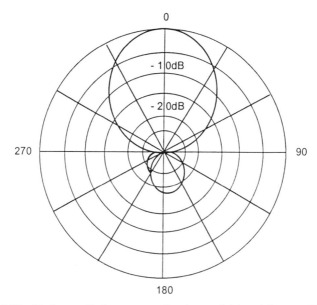

Figure 4.26   *E-plane radiation pattern of a short-axial-length horn at 200 MHz (after Kerr)*

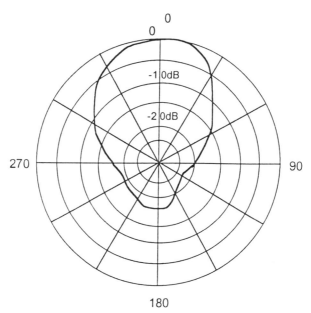

Figure 4.27   *H-plane radiation pattern of a short-axial-length horn at 200 MHz (after Kerr)*

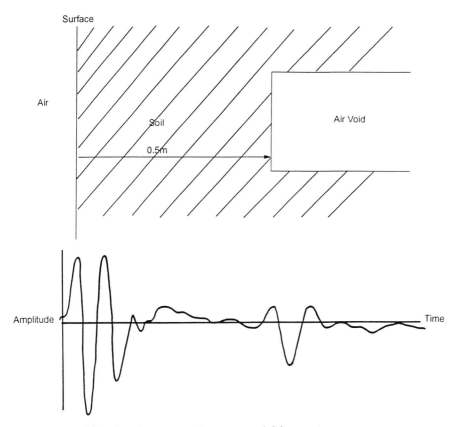

Figure 4.28   Time-domain response of an exponential horn antenna

ridged-horn design provides a characteristic impedance in the TE_{10} mode which is given by

$$z = Z_0 \, e^{kx} \text{ for } 0 \leqslant x \leqslant \frac{l}{2} \qquad (4.28)$$

where

Z_0 = characteristic impedance of the waveguide
k = average of start-and end-point impedances.

Experimentally, Walton and Sunberg (1964) found that the width of the ridge should be increased in the flared section to avoid the excessive phase errors and the resulting loss of gain which occurs when the ridge is near the aperture.

It is normal to use a phase-correcting lens, as typically the phase error is a quadratic function of the maximum aperture dimensions. Either a doubly

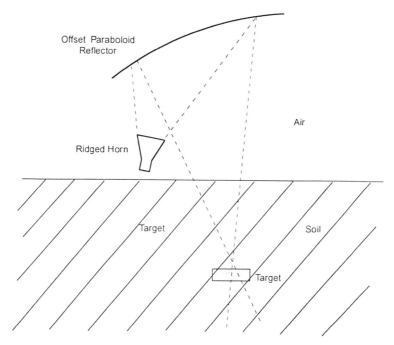

Figure 4.29   Arrangement of an offset-fed paraboloid reflector

planoconvex lens fed on the convex side can be used or a composite lens (one for each plane) can be employed.

Very often the horn antenna is used to feed an offset-fed parabolic reflector. For time domain systems the reflector will introduce transient delays and the transient time of a paraboloidal reflector, as shown in Figure 4.29, is given by (Sun and Rusch, 1991)

$$I_t = \frac{1}{c}\{D\sin\theta\cos(\phi_0 - \phi_m) + 2d\sin\theta_{\max}\cos(\phi_0 - \phi_m)\} \qquad (4.29)$$

Lai and Sinopoli (1992) obtained improvements in the performance of the ridged horn by the use of a combination of Vivaldi exponential slotline fed folded bowtie and rolled termination as shown in Figure 4.30.

The rolled termination, (which is also loaded with absorber), is used to minimise edge diffraction and well controlled radiation patterns can be produced over several octaves for a 70 cm 30° flare and plate angle antenna. Crosspolar levels are typically 20 dB below copolarised signal levels and a VSWR of better than 2:1 over at least three octaves can be achieved.

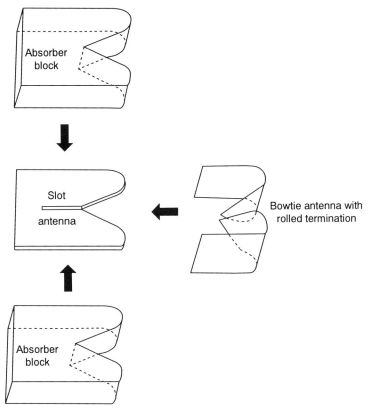

Figure 4.30   Composite-horn antenna (after Lai)

4.6 Array antennas

Future developments in surface-penetrating radar may be limited to the use of arrays of antennas. In the case of road survey the benefits of covering one lane of a carriageway using an array are obvious in terms of speed of survey. However, the ability to carry out beam forming by means of inverse synthetic aperture processing could be potentially valuable in many applications.

Rutledge and Muha (1982) consider the general situation of imaging antenna arrays while Anderson *et al.* (1991) consider the specific problem of wideband beam patterns from sparse arrays.

If an array of emitters is driven by a sequence of impulses without any differential time delay, the radiated time sequence is as shown in Fig. 4.31. Note the gradual disappearance of the sidelobes as radiated wavefront propagates away from the array. If the sequences of impulses are controlled in time by means of a differential time delay between each element, the beam position can be steered as shown in Figure 4.32.

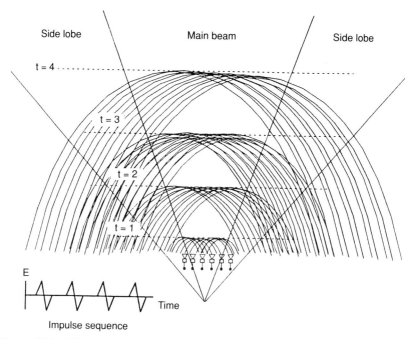

Figure 4.31   Timed-array antenna

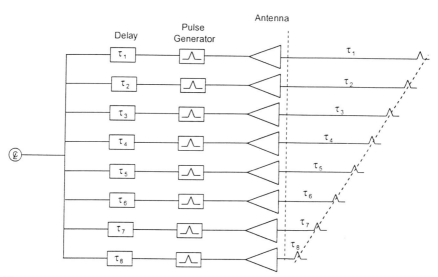

Figure 4.32   Beam steering by differential time delay

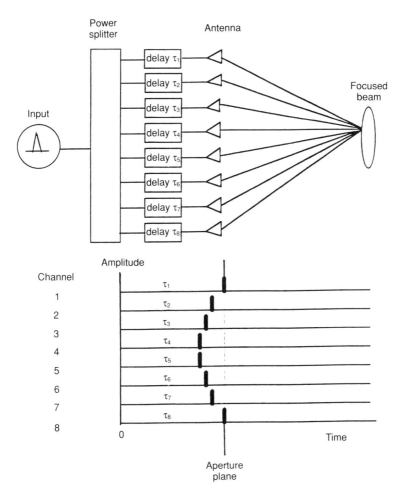

Figure 4.33   Beam forming by differential time delay

The possibility of beam steering by means of time control exists although the inter-element time delay is limited to a maximum equivalent to the distance between each element.

An alternative means of beam forming is by means of an array as shown in Figure 4.33. Here the objective was to create directivity in the azimuth plane and was achieved by means of a beam-forming network using wideband hybrid elements.

The use of time transmitter arrays and inverse synthetic-aperture processing of the signals from a receiver array offers the possibility of achieving considerably improved directivity over all the previous types of antennas discussed in this Chapter. A 10 × 10 element array will have a peak sidelobe amplitude of −26 dB of the main lobe (Anderson *et al.*, 1991) and beam steering of up to 50° is feasible.

4.7 Polarisation

It is well known that linear targets such as small-to-medium-diameter pipes act as depolarising features and a linearly polarised crossed-dipole antenna rotated about an axis normal to the pipe produces a sinusoidal variation in received signal. However, the null points are a distinct disadvantage, because the operator is required to make two separate, axially rotated measurements at every point to be sure of detecting pipes at unknown orientations, as shown in Figure 4.34.

An attractive technique is to radiate a circularly polarised wave which automatically rotates the polarised vector in space and hence removes the direction of signal nulls. Conventionally, circular polarisation refers to a steady-state condition during which a long duration pulse or CW waveforms are transmitted. For impulse radars, the pulse duration is very short (<5 ns) and hence a more complex transient situation is encountered.

One method of radiating circular polarisation is to use an equi-angular spiral antenna. Unfortunately, the dispersive nature of this type of antenna causes an increase in the duration of the transmitted waveforms and the radiated pulse takes the form of a 'chirp' in which high frequencies are radiated first, followed by the low frequencies. This effect, however, may be compensated by a 'spiking'

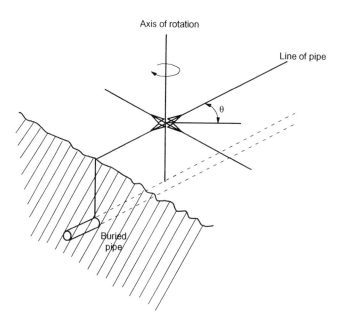

Figure 4.34   Crossed-dipole measurement on a linear target
$$V_r = V_t K \sin 2\theta$$

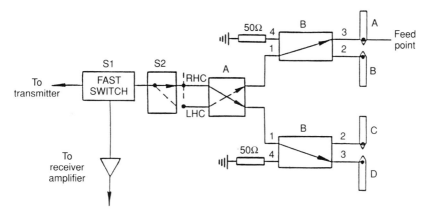

Figure 4.35 Synthesised-polarisation arrangement
      A = 90° wideband hybrid
      B = 180° wideband hybrid

filter, which may take the form of a conventional matched filter or a more sophisticated filter such as Wiener filter. The use of such an antenna has been shown by British Gas (Scott and Gunton, 1987) to be a useful method of implementing a pipe-detection radar and plastics pipes buried in wet clay have been detected up to a depth of 1.0 m.

A spiral antenna could be implemented as a multi-element planar structure as shown in Figure 4.20. However, this realisation may often fail to provide the expected performance because of several deficiencies. First, the limited physical dimensions of the spiral result in elliptical polarisation at low frequencies which can degrade to essentially linear polarisation; secondly reflection from the ends of the arms cause both clutter and degradation of the circularity of the polarisation and, thirdly, the proximity of the ground affects the reactive field of the antenna, resulting in nonsymmetrical loading and degradation of the far field pattern. An analysis of the type of antenna is given by Anders (1971).

A plot of axial ratio measured in free space as a function of angle at a frequency of 500 MHz is shown in Figure 4.22 and shows an acceptable performance of 1 dB. However, at low frequencies the axial ratio degrades quite rapidly reaching unacceptable values. Similarly with commercially available baluns the high frequency axial ratio degrades. The situation has a significant effect on the envelope of the polarisation vectors of the transmitted pulse. The consequence of these deficiences is that multiple angular measurements must be carried out, thus losing the original benefit of employing circular polarisation.

An alternative design possibility is to synthesise a circularly polarised signal. Any steady-state wave of arbitrary polarisation can be synthesised from two waves orthogonally polarised to each other. In the design shown in Figure 4.35, a circularly polarised wave is produced by exciting vertically and horizontally

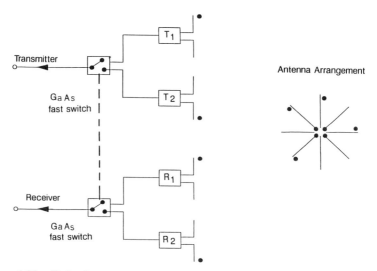

Figure 4.36   Eight-element commutated antenna
Note that system switches every alternate waveform

polarised waves, each having the same amplitude and with a 90° phase difference between them.

The radiating elements are fed, via wideband (preferably decade) 180° and 90° hybrids, to radiate circular polarisation. If right-hand circularly polarised signals are transmitted and received, the preferential detection of linear features (e.g. pipes) is achieved. If, however, right-hand circularly polarised signals are transmitted and left-hand circularly polarised signals are received, planar features are detected. Hence, if connections to the radiating elements are arrranged and switched appropriately, the signals routed to the receiver contain different data according to the sense of polarisation. The data, therefore, can be processed and in a different manner in order to provide images of different targets in the material under investigation.

However, hardware deficiencies limit the performance; first, it is difficult to achieve wideband operation with 90° hybrids (at least over a decade) and, secondly, even the fastest, state-of-the-art, GaAs switches have unacceptably high break-through levels.

An alternative concept is that of the commutated multi-element crossed-dipole array.

An example of an eight-element antenna in which the crossed-dipole pairs can be switched at intervals of up to 1 ms so that the two crossed-dipole pairs are orientated between 0° and 45° s shown in Figure 4.36.

An extension of this design concept is where full commutation over 360° in 45° steps can be achieved, and this is shown in Figure 4.37. PIN diode switches are used to handle the transmitted power and operate at a switching interval of 1 μs, thus achieving 360° rotation in approximately 10 μs. The possibility of real-time discrimination using filters based on recognition of the $\cos 2\theta$ amplitude

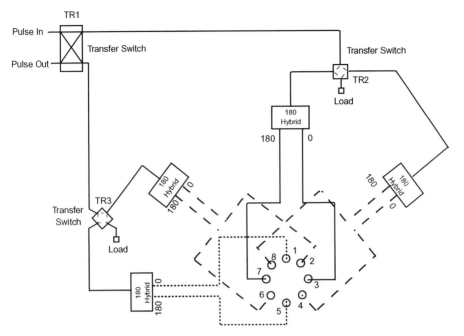

Figure 4.37   Fully commutated antenna arrangement

variation of each range sample can be considered. Operationally, such a system would have the advantage of being able to survey rapidly without the limitations imposed by mechanically rotated antennas.

However, all of these schemes do not solve the problem of detecting deeply buried, large-diameter pipes (> 250 mm) for which the depolarisation of incident signals is low. In this case it may be necessary to detect the copolar signal and a parallel-dipole-antenna system may be needed.

Antenna-design approaches based on commutated crossed dipoles, as described in this Section, could prove a successful means of detecting plastic pipes up to 250 mm diameter, provided that the radar system has sufficient dynamic range and sensitivity.

4.8 Summary

The antennas used in surface-penetrating-radar systems are, for reasons of portability, usually electrically small and consequently exhibit low gain. This has a profound effect on the performance of the overall system and is probably the only example of a radar system where antenna gain is, in general, so low. However, the bandwidth of the antennas is very much greater than that normally used in conventional radar systems, and surface-penetrating radars generally demonstrate very high range resolution.

The choice of antenna is generally straightforward. The resistively loaded dipole, bowtie and TEM travelling-wave antenna have been primarily used for the impulse-based radar. Where matched filtering can be incorporated in impulse radars, then either horn or frequency-independent antennas can also be considered.

All the classes of antenna discussed can be used in synthesised, FMCW or noise-modulated radars.

Attention must also be given to the means by which the antenna is fed from the transmitter. Generally an antenna is a balanced structure but where cables are used to connect the antenna to the transmitter or receiver, some means is needed of transforming from the unbalanced configuration of the feed cable to the balanced structure of antenna. On this frequency range, baluns are generally commercially available or, alternatively, purpose-designed units can be constructed.

It may be found that multiple reflections between the transmitter and antenna can be troublesome and these can be avoided by making the feed cable long enough to place the reflection outside the time window of interest. This, however, may be undesirable as the cable will act as a lowpass filter unless compensated. An alternative is to mount the transmitter and receiver immediately adjacent to the antenna and this, if correctly designed, can remove the need for either a balun or feed cable.

Artificial dielectric loading has been used with antennas for surface-penetrating radar. However, the finite size of the dielectric can cause problems in that energy becomes trapped in the dielectric and has the effect of causing a local resonance. Provided, therefore, that the antenna can be loaded in such a way that the dielectric appears semi-infinite this problem is reduced.

The expected main developments in the field of antennas appear to be related to array antennas. The current interest in the development of free-space ultra-wideband radar systems may result in the transfer of useful developments.

4.9 References

ALTSHULER, E. E. (1961): The traveling-wave linear antenna. *IRE Trans.*, **AP–9**, (4), 324–329

ANDERS, R. (1971): On the complete calculation of the equiangular spiral antenna. Proceedings of the 20th international symposium on *Antennas and Propagation*, Japan, 19–20

ANDERSON, F., FULLERTON, L., CHRISTENSEN, W., and KORTEGAARD, B. (1991): Wideband beam patterns from sparse arrays. Proceedings of the first Los Alamos symposium on *Ultra-Wideband Radar*. 1990, CRC Press

BAWER, R., and WOLFE, J. J. (1960): The spiral antenna. IRE National Convention Record, Part J, 84–95

BREWITT-TAYLOR, C. R., GUNTON, D. J., and REES, H. D. (1981): Planar antennas on a dielectric surface. *Electron. Lett.*, **17**, (20), 729–731

BROWN, G. H., and WOODWARD, O. M., Jun. (1952): Experimentally determined radiation characteristics of conical and triangular antennas. *RCA Rev.*, **13**, 425–452

BURKE, G. J., JOHNSON, W. A., and MILLER, E. K. (1983): Modeling of simple antennas near to and penetrating an interface. *Proc. IEEE*, **71**, (1), 174–175

DANIELS, D. J. (1980): Short pulse radar for stratified lossy dielectric layer measurement. *IEE Proc. F*, **127**, (5), 384–388

DESCHAMPS, G. (1959): Impedance properties of complementary multiterminal planar structures. *IRE Trans.*, **AP–7**, 5371–5378

DYSON, J. D. (1959): The equiangular spiral antenna. *IRE Trans.*, **AP–7**, 181–187

ESSELLE, K. P., and STUCHLY, S. S. (1990): A new broadband antenna for transient electromagnetic field measurements. Proceedings of IEEE symposium on *Antennas and Propagation – Merging Technologies for 90s*, Dallas, TX, USA, 1584–1587

EVANS, S., and KONG, F. N. (1993): TEM horn antenna: input reflection characteristics in transmission. *IEE Proc. H*, **130**, (6), 403–409

FOSTER, P. R., and TUN, S. M. (1993): Design and test of two TEM horns for ultrawideband use. IEE colloquium on *Antennas and propagation problems of ultrawideband radar*, *IEE Colloquium Digest* 004

GIBSON, P. J. (1979): The Vivaldi aerial. Proceedings of the seventh *European Microwave Conference*. Microwave Exhibitions & Publishers Ltd., 101–105

GOLDSTONE, L. L. (1983): Termination of a spiral antenna. *IBM Tech. Disclosure Bull.*, **25**, (11A), 5714–5715

HARMUTH, H. F. (1983): Antennas for nonsinusoidal waves. Part III Arrays. *IEEE Trans.*, **EMC-25**, (3), 346–357

HARMUTH, H. F., and DING-RONG, S. (1983*a*): Antennas for nonsinusoidal waves. Part II Sensors. *IEEE Trans.*, **EMC–25**, (2), 107–115

HARMUTH, H. F., and DING-RONG, S. (1983*b*): Large–current, short-length radiator for nonsinusoidal waves. Proceedings of IEEE international symposium on *Electromagnetic Compatibility*, Arlington, TX, USA, 453–456

ILZUKA, K. (1967): The traveling-wave V-antenna and related antennas. *IEEE Trans.*, **AP–15**, (2), 236–243

JUNKIN, G., and ANDERSON, A. P. (1988): Limitations in microwave holographic imaging over a lossy half space. *IEE Proc. F*, **135**, (4), 321–329

KAPPEN, F. W., and MÖNICH, G. (1987): Single pulse radiation from a resistive coated dipole-antenna. Proceedings of fifth international conference on *Antennas and Propagation*, ICAP '87, York, UK, *IEE Conf. Publ.* 274, Vol 1, 90–93

KERR, J. L. (1973): Short axial length broad-band horns. *IEEE Trans.*, **AP–21**, 710–714

KING, R. W. P., and NU, T. T. (1993): The propagation of a radar pulse in sea water. *J. Appl. Phys.*, **73**, (4), 1581–1589

KING, R. W. P., and SMITH, G. S. (1981): *Antennas in matter*, MIT Press

KOOY, C. (1984): Impulse response of a planar sheath equi-angular spiral antenna. *Arch. Elektron. Uebertrag.tech.*, **38**, (2), 89–92

LAI, A. K. Y., and SINOPOLI, L. (1992): A novel antenna for ultra-wide-band applications. *IEEE Trans.*, **AP–40**, (7), 755–760

MALONEY, J. G., and SMITH, G. S. (1991): The role of resistance in broadband, pulse-distortionless antennas. Proceedings of IEEE Antennas & Propagation Society symposium, 24–28 June 1991, Vol. 2, 707–710

MILLER, E. K., and LANDT, J. A. (1977): Short-pulse characteristics of the conical spiral antenna. *IEEE Trans.*, **AP–25**, (5), 621–626

MORGAN, T. E. (1985): Spiral antennas for ESM. *IEE Proc. F*, **132**, (4), 245–251

PASTOL, Y., ARSAVALINGAM, G., and HALBOUT, J.-M. (1990): Transient radiation properties of an integrated equiangular spiral antenna. Proceedings of IEEE symposium on *Antennas and Propagation — Merging Technologies for 90s*, Dallas, TX, USA, 1934–1937

RANDA, J., KANDA, M., and ORR, R. D. (1991): Resistively–tapered–dipole electric–field probes up to 40 GHz. Proceedings of IEEE international symposium on *Electromagnetic Compatibility*, Cherry Hill, NJ, USA, 265-266

RAO, B. R. (1991): Optimized tapered resistivity profiles for wideband HF monopole antenna. Proceedings of the IEEE Antennas & Propagation Society symposium, Vol. 2, 711–713

READER, H. C., EVANS, S., and YOUNG, W. K. (1985): Illumination of a rectangular slot radiator over a 3 octave bandwidth. Proceedings of fourth international conference on *Antennas and Propagation*, ICAP '85, Coventry, UK, *IEE Conf. Publ.* 248, 223–226

RUMSEY, V. H. (1966): *Frequency independent antennas*. Electrical Science Series, Academic Press

RUTLEDGE, D. B., and MUHA, M. S. (1982): Imaging antenna arrays. *IEEE Trans.*, **AP–30**, (4), 535–540

SCOTT, H. F., and GUNTON, D. J. (1987): Radar detection of buried pipes and cables. Institution of Gas Engineers, 53rd Annual meeting, London, UK, Communication 1345

SHEN, H. M., KING, R. W. P., and WU, T. T. (1988): V-conical antenna. *IEEE Trans.*, **AP-36**, (11), 1519–1525

SKOLNIK, M. I. (1970): *Radar handbook*. McGraw-Hill, 9–6 to 9–11

SUN EN-YUAN and RUSCH, W. V. T. (1991): Transient analysis of large reflector antennas under pulse-type excitation. Proceedings of IEEE Antennas & Propagation Society symposium, 674-677

THEODOROU, E. A., GORMAN, M. R., RIGG, R. R., and KONG, F. N. (1981): Broadband pulse-optimised antenna. *IEE Proc. H*, **128**, (3), 124–130

WAIT, J. R. (1960): Propagation of electromagnetic pulses in a homogeneous conducting earth. *Appl. Sci. Res. B*, **8**, 213–253

WALTON, K. L., and SUNBERG, V. C. (1964): Broadband ridged horn design. *Microw. J.*, 96–101

WOHLERS, R. J. (1970): The GWIA an extremely wide bandwidth low dispersion antenna. Abstracts of 20th symposium, US Air Force R&D Programme

WU, T. T., and KING, R. W. P. (1965): The cylindrical antenna with nonreflecting resistive loading. *IEEE Trans.*, **AP-13**, (5), 369–373

YOUNG, M., *et al.* (1977): Underground pipe detector. US patent 4 062 010

Chapter 5

Modulation techniques

5.1 Introduction

Each of the various modulation techniques used for surface-penetrating-radar systems has its relative merits and in this chapter we consider the general system architecture and system specifications associated with each.

The most frequently used system design is that of the impulse radar and the majority of commercially available radar systems use short pulses or impulses which generally come in the category of amplitude modulation (AM). The next most frequently used modulation technique is frequency modulation (FM) followed by synthesised pulse (SPM), holographic (HM) and finally coded and noise modulation (NM). In this Chapter we consider the most commonly used techniques in turn and discuss the key parameters which need to be considered in the design process.

It is useful to state the relationship between a time-domain function $f(t)$ and its representation in the frequency domain $f(\omega)$. Fourier transform methods can be used to show that

$$F\{f(t)\} = f(\omega) \tag{5.1}$$

that is

$$f(\omega) = \frac{1}{2\pi} \int_{-\infty}^{\infty} f(t) \, e^{j\omega t} dt \tag{5.2}$$

The fast Fourier transform (FFT) is often used to compute the transform but certain well established restrictions must be borne in mind owing to its finite limitations. Here we shall consider only the results of applying the Fourier transform over infinite limits.

A real function such as a monocycle is given by the expression

$$f(t) = A\sin(\omega' t) \text{ over the range } -\frac{\tau}{2} < t < \frac{\tau}{2}$$

$$\tag{5.3}$$

$$f(t) = 0 \text{ elsewhere}$$

The Fourier transform of $f(t)$ is given by

$$F(\omega) = \frac{1}{2\pi} \int_{-\frac{T}{2}}^{\frac{T}{2}} A \sin \omega' t \; e^{-j\omega t} dt \tag{5.4}$$

that is

$$f(\omega) = \frac{1}{2\pi} [\text{sinc } (\omega' + \omega) T - \text{ sinc } (\omega' - \omega).T \tag{5.5}$$

and takes the form shown in Figure 5.1. From this it can be seen that the required bandwidth is significant. However, in most cases the material acts as a lowpass filter and attenuates the higher frequencies. Thus the system designer must therefore consider the effect of the material on system performance as the general effect is to distort the received time-domain waveform by extending its duration and reducing the bandwidth of received information.

Faithful transmission of the particular modulated wave imposes certain requirements on the hardware used in each type of system and we shall consider each of these in turn in the following Sections. Considerable bandwidth is required to achieve adequate levels of resolution for a surface-penetrating radar and the resolution equivalent to a monocycle dictates a bandwidth of the order of three to five octaves. This is a significant requirement and increases the cost and complexity of the transmitter–receiver system. Evidently as the bandwidth of the system decreases, the effect is to increase the duration of the radiated/received signal; hence a compromise is reached between desired resolution/system bandwidth and the characteristics of the material being probed.

The following Sections examine the different classes of modulation and the general system architecture and hardware used to construct a suitable radar system. In general, the output from most systems results in an equivalent time-domain representation of the waveform and hence subsequent signal-processing methods will be referred to only where system architecture is different, as for example for a holographic imaging radar.

5.2 Amplitude modulation

The majority of surface-penetrating-radar systems have used impulses of radio frequency energy variously described as baseband, video, carrierless, impulse, monocycle or polycycle. The simplified general block diagram of an amplitude modulated system is shown in Figure 5.2, and a timing diagram shown in Figure 5.3.

A sequence of pulses typically of amplitude within the range 20–200 V and pulse width within the range 200–50 ns at a pulse-repetition interval of between several hundred microseconds and one microsecond, depending on the system

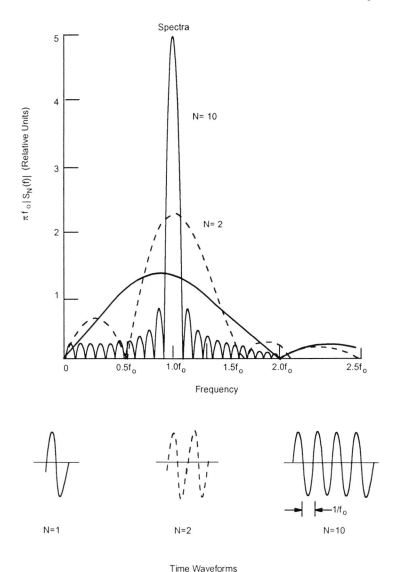

Figure 5.1   *Spectrum of a single-cycle waveform compared with the spectra of two-cycle and ten-cycle waveforms (after Skolnik)*

design, is applied to the transmit antenna, although it is quite feasible to generate pulses of several hundred kilovolts albeit at long repetition intervals. The output from the receive antenna is applied to a flash analogue–digital converter or a sequential sampling receiver. This normally consists of an ultra-high-speed sample-and-hold circuit. The control signal to the sample-and-hold circuit which determines the instant of sample time is sequentially incremented each pulse-repetition interval. For example, a sampling increment of $t' = 100$ ps

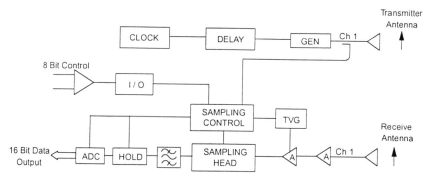

Figure 5.2   Transmitter–receiver architecture

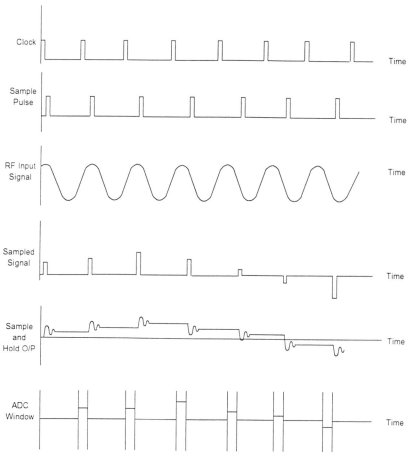

Figure 5.3   Sequential-sampling-receiver timing

is added to the previous pulse-repetition sampling interval to permit sampling of the received signal at regular intervals as indicated below:

$$T_s = T + nt' \text{ for } n = 1 \text{ to } N \tag{5.6}$$

$$T = \text{pulse repetition time}$$

$$t' = \text{sampling interval}$$

$$N = \text{total number of samples}$$

Certain important limitations in terms of sampling interval should be noted. From the sampling theorem, the sampling interval must be such as to comply with the Nyquist relationship

$$t' < \frac{1}{(2B)} \tag{5.7}$$

where B is the bandwidth of interest. In practice, a greater number of samples is normally required for accurate reconstruction and the sampling interval is generally taken as

$$t' \cong \frac{1}{5B} \tag{5.8}$$

The principle of the sampling receiver is therefore a down conversion of the radio-frequency signal in the nanosecond time region to an equivalent version in the micro- or milli-second time region. The incrementation of the sampling interval is terminated at a stage when, for example, 256, 512 and 1024 sequential samples have been gathered. The process is then repeated. There are several methods of averaging or 'stacking' the data; either a complete set of samples can be gathered and stored and further sets added to the stored data set or alternatively the sampling interval is held constant for a predetermined time to accumulate and average a given number of individual samples. The first method needs a digital store but has the advantage that each waveform set suffers little distortion if the radar is moving over the ground.

The second method does not need a digital store and a simple lowpass analogue filter can be used. However, depending on the number of samples that have been averaged, the overall waveform set can be 'smeared' spatially if the radar is moving at any speed.

The stability of the time increment is very important and generally this should be 10% of the sampling increment; however, in practice a stability of the order of 10–50 ps is achieved. The effect of timing instability is to cause a distortion which is related to the rate of change of the RF waveform. Evidently where the RF waveform is changing rapidly, jitter in the sampling circuits results in a very noisy reconstructed waveform. Where the rate of change of signal is slow, jitter is less noticeable. Control of the sampling converter is normally derived from a

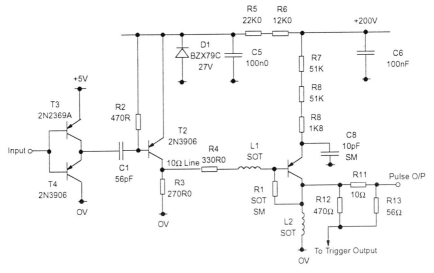

Figure 5.4   Avalanche-transistor impulse generator

sample of the output from the pulse generator to ensure that variations in the timing of the latter are compensated for automatically.

The key elements of this type of radar system are the impulse generator, the timing control circuits, the sampling detector and the peak hold and analogue–digital converter.

The impulse generator is generally based on the technique of rapid discharge of the stored energy in a short transmission line. The most common method of achieving this is by means of a transistor operated in avalanche breakdown mode used as the fast switch and a very short length of transmission line. A typical circuit arrangement is shown in Figure 5.4 and this provides an output of 100 V with a duration of 1 ns as shown in Figure 5.5. The frequency-domain characteristics of such an impulse are shown in Figure 5.6. If shorter-duration

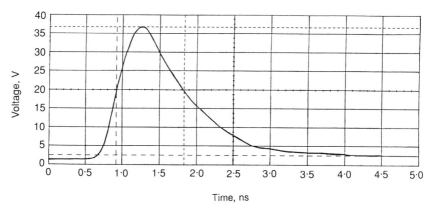

Figure 5.5   Typical voltage waveform generated by the circuit of Figure 5.4

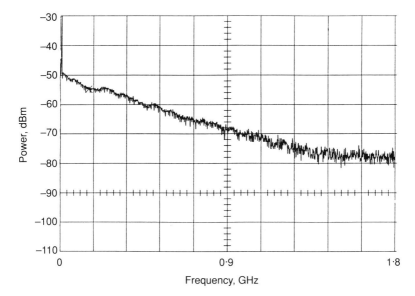

Figure 5.6   Typical spectrum of the waveform generated by the circuit of Figure 5.4

impulses are required it is possible to use a step-recovery diode to generate impulses of durations of the order of 200 ps and a typical output voltage is in the region of 30 V.

It is evident that the repetitive nature of the pulse causes a line spectrum in the frequency domain. The typical repetition time interval for an avalanche-transistor impulse generator is of the order of 0.1–10 μs, while the step-recovery diode must be chosen specifically to match the repetition interval to ensure that charge-carrier re-combination can take place.

Other methods of generating impulses use power FETs and voltage impulses up to 10 kV have been generated (Hares, 1992). An alternative means of generating high-power impulses is based on the use of a photoconductive semiconductor switch (PCSS) to discharge a capacitor into a short-circuited transmission line. A picosecond laser pulse is used rapidly to switch the conducting/nonconducting state of a semiconductive material such as GaAs. Typical output voltages of 14 kV in a 50 Ω impedance for a duration of nanoseconds or less have been produced (Pocha *et al.*, 1989).

A further variation on this technique is the frozen-wave generator shown in Figure 5.7. This consists of several segments of transmission lines connected in series by means of picosecond photoconductive switches. The output from the ensemble is a sequential waveform of arbitrary characteristic, i.e. a frozen wave'. Output voltage in the kilovolt range has been generated (Lee, 1991).

Several factors need to be considered in the design of impulse sources: reliability, jitter and repetition rate. For avalanche devices the avalanche process is statistical by nature and is accompanied by jitter. For optical devices the physics of the device must be considered, as the lifetime of the carriers

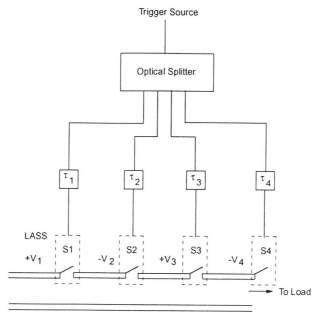

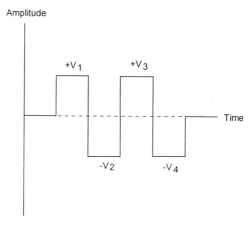

Figure 5.7   *'Frozen-wave' pulse generator*
            LASS = light activated silicon switch

determines the recombination time of the material and, for silicon, where $T \cong 100~\mu s$, may restrict the repetition frequency of the switch. GaAs, on the other hand, exhibits a recombination time of 1 ns. Optical switches may exhibit a reliability of up to 10^8 operations which means that their lifetime can be significantly reduced by operation of the radar at high-pulse-repetition rates.

The high-speed-sampling approach conventionally used to display fast waveforms produces a low signal/noise ratio because the spectrum of the sampling pulse is a poor match for that of the received pulse. Being an essentially nonselective filter, it allows large amounts of noise energy to enter the receiver. Also, the sampling circuit tends to add milliampere-level unbalanced currents as well as sampling pulse noise to its output. Although an acceptable trade off for usual laboratory purposes, this may be unacceptable for receivers with sensitivity in the microvolt range.

Alternative methods of data acquisition are based on high-speed analogue–digital converters or the crosscorrelator receiver. There are several methods of acquiring the high-bandwidth RF signals output from the receiver; direct analogue–digital conversion using high-speed (flash) converters, frequency selection followed by high speed analogue–digital conversion, or sequential sampling.

Typical flash analogue–digital converters feature large signal bandwidths of many hundreds of megahertz, sampling jitter less than 5 ps and 8-bit resolution. At bandwidths over 500 MHz, typical sampling resolutions are 4 bit and a more complex system architecture is found. In general, most of the current generation of impulse radars use high-speed analogue–digital conversion receivers for bandwidths below 200 MHz where greater resolution can be achieved.

An alternative receiver architecture is based on subdividing the RF frequency band and mixing the individual band to separate intermediate-frequency bandwidths of 200 MHz that can be then separately analogue–digital sampled.

The wideband crosscorrelator receiver can use coherent processing and, if required, time-dithered decoding. The crosscorrelator is equivalent to a matched filter but has more flexibility. The reference waveform can be matched to the transmitted waveform, or for radar, to the complex signature of a particular target, and can be changed in real time if needed. The components required to construct the crosscorrelator are small, inexpensive, of low power and compatible with VLSI techniques. It is thus possible to have a large number of correlators operating independently in parallel to achieve the throughput necessary to provide high resolution, real-time, time-coded operation.

Sampling, as referred to above, is a time-extending process with which a high-frequency repetitive signal is duplicated at a lower repetition rate. This type of sampling, where each sample is taken at a fixed frequency with the period of time between samples remaining constant, is known as coherent sampling.

The most basic sampling gate is a simple single diode as shown in Figure 5.8. The diode is essentially a switch, normally 'open' (diode reverse-biased). A short pulse momentarily closes the switch (diode forward-biased) allowing charge to flow from the source to be stored in the capacitor C_s, which results in a voltage across C_s proportional to the input signal. The pulse width must be narrow compared with the period of the input signal so that the sample corresponds to a specific portion of the input waveform. The capacitor-charging time must be fast enough to accept the charge during this pulse time.

The problems of isolation between the signal circuit and the sampling pulse

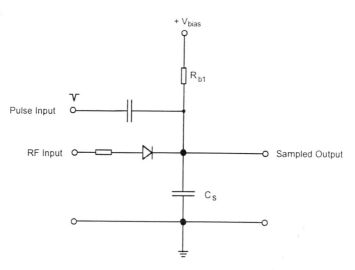

Figure 5.8   Single-diode sampling gate

and bias circuits can be serious with the single-diode sampler. A two-diode sampler, however (shown in Figure 5.9), has a low sampling efficiency. The efficiency can be improved by substituting two more diodes for the two resistors in the bridge.

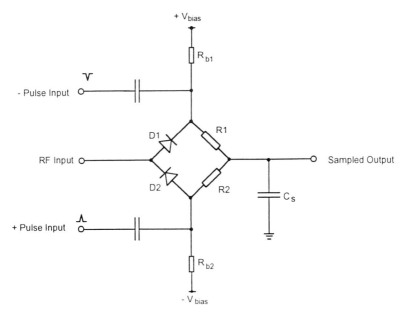

Figure 5.9   Dual-diode sampling gate

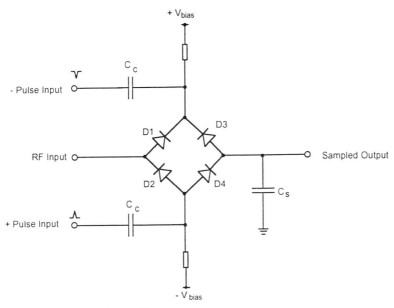

Figure 5.10   Quad-diode sampling gate

The four diode sampling gate shown in Figure 5.10 is the most commonly used. In a sampling system it would be situated between the input source and the input capacitor of an amplifier. The diodes are normally reverse biased so that the input signal does not cause them to conduct. Sampling is initiated with very narrow pulses which overcome the reverse bias and switch the diodes into conduction. The low-impedance paths allow the amplifier input capacitor to be charged to a voltage proportional to the input voltage. Owing to the short charging time, the capacitor may not be charged to the full input voltage. (Some systems include feedback control circuits to continue charging the capacitor in between pulses until the capacitor voltage equals the input voltage.) The capacitor remains charged until the next pulse.

The reverse bias applied to the sampling-gate diodes is a critical factor in the operation of a sampler. It must be large enough to prevent input signals driving the diodes into conduction and small enough to allow the gating pulses to forward bias the diodes during the sampling periods to achieve maximum sampling efficiency.

Both DC and AC balance of the sampling-gate bridge are essential in achieving the symmetry required for optimum performance of the sampler. The conditions of balance require that the four sampling diodes be matched, the two reverse-bias voltages be equal and opposite, and the sampling-gate control voltage be identical in waveshape except for polarity. One method of providing identical control-gate signals is to derive them from the identical and bifilar-wound windings of a transformer. A narrow pulse can be produced as a result of differentiation of a rectangular pulse with a small coupling capacitor. If

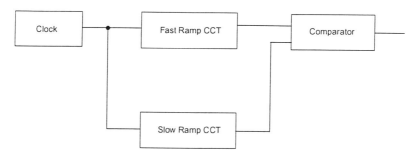

Figure 5.11   Dual-ramp timing circuit

required, pulse risetimes can be reduced by using step-recovery diodes before coupling through the capacitor.

The timing-control circuits are a key element of the receiver, and the standard method of generating a sequence of incremented pulses is by means of a dual ramp circuit as shown in Figure 5.11. The fast ramp is at the same rate as the pulse-repetition time while the slow ramp is set to provide the desired number of samples, i.e. 256, 512 or 1024. The timing sequence for this is shown in Figure 5.12.

The ramp circuits can be designed using analogue or digital circuits. When analog circuits are used the main building block is an integrator circuit whereas when a digital design is adopted suitable integrated circuits are available in the form of Analog Devices AD9500 digital-delay IC.

Evidently, time stability of these circuits is vitally important and, for example, a 512 ns time window with 256 samples requires one sample every 200 ps increment. However, this increment occurs every 1 μs; hence timing stability must be $1\mu s + 200$ ps $\pm$ 20 ps, i.e. 1000.2 ns $\pm$ 20 ps or $\pm$ 0.002%. Great care is therefore needed in circuit design to achieve adequate stability.

The dynamic performance of an impulse (amplitude modulated) radar can be estimated from the following example, which considers a system with the characteristics given in Table 5.1.

A plot of received signal strength against time is shown in Figure 5.13. From this it can be seen that the operating range of the radar system lies between the boundaries defined by the functions defined by the clutter profile, target reflection loss and the limit of sensitivity due to the noise figure of the receiver.

It can be readily appreciated that the limited dynamic range of the sampling receiver limits the performance of the radar.

The poor noise figure of the sampling gate can be improved by using a wideband low-noise RF amplifier prior to the gate. The typical noise figure of a 1 GHz amplifier is 2.4 dB; hence an immediate improvement in system noise figure is achieved. However, the sampling gate may now be vulnerable to saturation by high-level signals caused by targets at very short ranges.

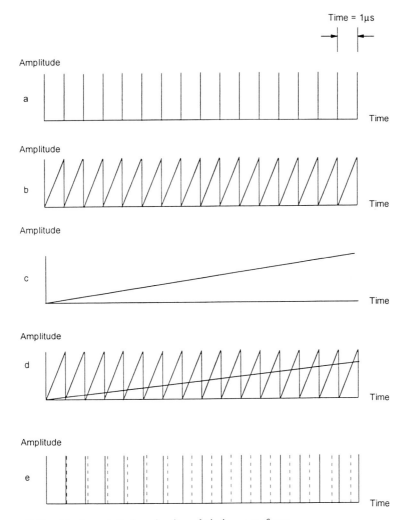

Figure 5.12 Dual-ramp timing circuit and timing waveforms
  a = Clockwaveform
  b = Fast-ramp waveform
  c = Slow-ramp waveform
  d = Inputs to comparator circuit
  e = Output from comparator circuit

The solution to this problem is to incorporate an additional RF amplifier whose gain can be varied as a function of time. In practice this is most easily achieved in synchronism with the pulse-repetition rate. This avoids undesirable intermodulation effects which can occur if the gain is changed in real time. This technique enables the receiver to be operated at maximum

Table 5.1   Typical characteristics for an impulse radar

Transmitter	Peak power	= 50W (46.9 dBm)
	Mean power	= 50 mW (16.98 dBm)
	Antenna and cable losses	= −16.9 dB
	Peak radiated power	= 1 W (30 dBm)
	Mean radiated power	= 1mW (0 dBm)
	Impulse duration	= 1 ns
	Impulse repetition time	= 1 μs
	System clutter profile (rate of decay of time sidelobes and cross coupling)	= 30 dB/ns
Receiver	RF Bandwidth	= 1 GHz
	Equivalent thermal noise (300 k)	= $4.14 \times 10^{-12} W(-84$ dBm)
	Noise figure of sampling head	= 40 dB
	Noise floor	= −44 dBm
	Maximum input signal level	= 7 dBm
	Dynamic range	= 53 dB

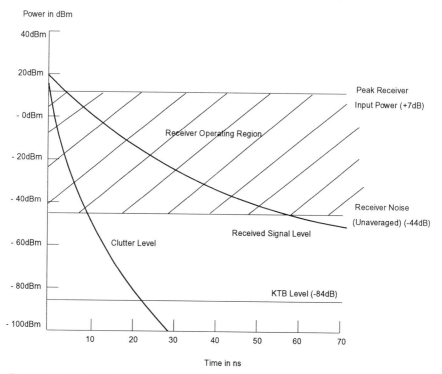

Figure 5.13   Received signal level against time for a sequential-sampling receiver

sensitivity without encountering overload problems. Ideally the gain/time characteristic should be related to the attenuation and reflection characteristics of the material under investigation. Hence an adaptive calibration method is advisable.

Generally a compression range of up to 40 dB can be expected and this is adequate to compress most of the high-level close-range signals. The optimum technique is to use an adaptive signal-level compression whereby the peak received signal as a function of time is adaptively set to a predetermined value by means of LNA (low noise amplifier) gain adjustment.

It is also possible to improve the dynamic range by averaging the received signal, and this improvement is given (in decibels) by

$$A = 20 \log \sqrt{N} \tag{5.9}$$

where N = number of averages.

However, the rate of improvement quickly reduces as N is increased and in practice 16 averages provide a reasonable improvement without excessive time penalties.

The frequency range of the output signal usually occupies a bandwidth up to 20 kHz. For a radar operating at a repetition rate of 1 μs with 256 samples each averaged 16 times, the ensemble downconverted signal is repeated at the time given by

$$\tau' = \tau_r N_a N_s \tag{5.10}$$

which, for the values given above, is equal to

$$\tau' = 1 \times 10^{-6} \times 16 \times 256$$
$$\tau' = 4 \text{ ms}$$

Hence the bandwidth of the downconverted signal is given by

$$B = \frac{1}{5\tau^*} \tag{5.11}$$

where τ^* = equivalent downconverted time per sample

$$B = \frac{1}{5\tau_r N_a}$$
$$B = 12\ 500 \text{Hz}$$

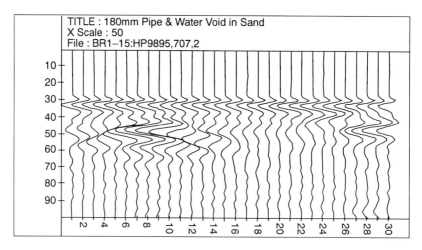

Figure 5.14   Example of time-domain signal received from two buried targets (courtesy ERA Technology)
Vertical scale = 10 units = 3 ns
Horizontal scale = 1 increment = 20 mm

Note that the true noise bandwidth of the receiver is defined by the RF bandwidth and the noise energy is converted together with the signal.

The dynamic range of the analogue–digital converter which follows the sampling head should be matched to the dynamic range of the latter, and typically a 12-bit or 16-bit converter is used.

When constructing impulse-radar systems it is necessary to ensure that adequate decoupling of the internal power supplies is achieved as the effect of impulsive noise from switched-mode power supplies on the sampling circuits can result in serious degradation of overall system performance. Hence good engineering practice must be maintained in the design and layout of the RF circuits.

It is also important to consider the physical layout of the sampling receiver, pulse generator and antennas. Two options are available. The antenna can be directly connected to the transmitter and receiver circuits or it can be interconnected via a length of high quality RF cable. In the latter case the physical length serves to separate electrically the reflected signals caused by the antenna and the transmitter/receiver circuits. However, the cable acts as a lowpass filter which degrades the system resolution. Where the antennas are directly connected, multiple echoes can prove difficult to reduce to low levels and some design skill is needed to achieve acceptable results. Additionally, the physical proximity of electronic components to the antennas may disturb their radiation characteristics.

Typical examples of time domain signals received from buried targets are shown in Figures 5.14 and 5.15.

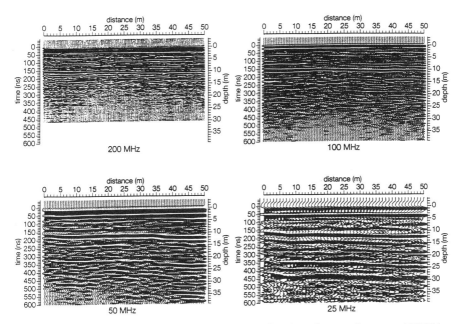

Figure 5.15   Example of time-domain signal as a function of centre frequency (TNO)

5.3 Frequency-modulated continuous-wave (FMCW) radar

Frequency-modulated continuous-wave (FMCW) radar systems have been used in preference to AM systems where the targets of interest are shallow and frequencies above 1 GHz can be used. As the centre frequency of operation increases, it is easier to design and build FMCW radars with wide bandwidths, whereas it becomes progressively more difficult to design AM systems. The block diagram of a typical FMCW radar is shown in Figure 5.16.

The main advantages of the FMCW radar are the wider dynamic range, lower noise figure and higher mean powers that can be radiated. In addition, a much wider class of antennas, i.e. horns, logarithmic elements etc., is available for use by the designer. In this Section we will consider continuously changing frequency modulation.

An FMCW radar system transmits a continuously changing carrier frequency by means of a voltage-controlled oscillator (VCO) over a chosen frequency range on a repetitive basis. The received signal is mixed with a sample of the transmitted waveform and results in a difference frequency which is related to the phase of the received signal and hence to its time delay and hence the range of the target. The difference frequency or intermediate frequency (IF) must be derived from an I/Q mixer pair, if the information equivalent to a time-domain representation is required, as a single-ended mixer only provides the modulus of the time-domain waveform.

In an FMCW radar, the transmitter frequency is changed as a function of

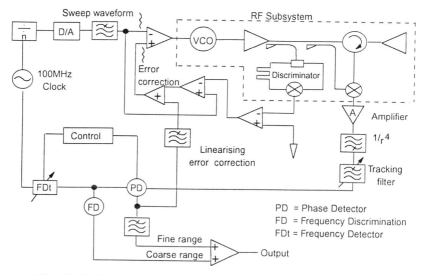

Figure 5.16   Block diagram of an FMCW radar system

time in a known manner. If the change is linear, then a target return will exist at a time T_d given by

$$T_d = 2R/c \tag{5.12}$$

where:

R = range in metres
c = velocity of light in metres per second

If this target return signal is mixed with the transmitted signal, a beat frequency, termed the IF (intermediate frequency), will be produced.

This will be a measure of the target range as shown in Figure 5.17. If the transmitted signal is modulated with a triangular modulating function at a rate f_m over a range f then the intermediate frequency is given by

$$f = \frac{4Rf_m\Delta f}{c} \text{ Hertz} \tag{5.13}$$

where:

f_m = modulation frequency in Hz
Δf = frequency deviation in Hz

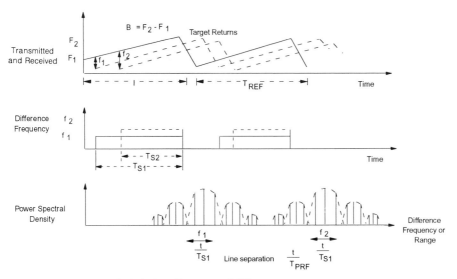

Figure 5.17 *FMCW timing diagram and IF system*
a RF transmitted and received waveforms
b Receiver output
c Frequency spectra of receiver output

The choice of modulating waveform defines the resultant IF spectrum and it is desirable to minimise the bandwidth of the IF spectrum due to a single target in order to optimise the target resolution. If the IF were generated by a continuous linear frequency deviation, then the IF spectrum would consist of sum and difference frequencies, a DC component and various other frequencies resulting from the mixer's nonlinear properties, assuming that the mixer had a perfect square-law characteristic. For practical purposes only the difference frequency will be considered.

However, the repetitive nature of the modulating waveform causes points in the IF time waveform where the amplitude drops to zero. This can be regarded as an amplitude modulation of the IF signal.

If the case of a single target is considered, the IF waveform as a function of time would be given by

$$f_{IF}(t) = A \sin \omega_{if} t \tag{5.14}$$

The repetitive nature of the RF sweep effectively convolves the basic IF waveform with the line spectra:

$$f_m^* = F\left(A \sin \omega_{if} t\right) * F\left(f_m^*(t)\right) \tag{5.15}$$

The Fourier transform of $f_{if}(t)$ is

$$F(A\sin\omega_{IF}t) = F\left[A\left(\frac{e^{j\omega_{if}t} - e^{-j\omega_{if}t}}{-j2}\right)\right]$$
$$= -A(j\pi\delta(\omega - \omega_{if}) + j\pi\delta(\omega + \omega_{if}) \quad (5.16)$$

The Fourier transform of $f_m^*(t)$ can be derived after rearranging the integration period:

$$f_m(t) = 1 \text{ for } 0 < t < (T - \tau) \qquad (5.17)$$
$$= 0 \text{ for } (T - \tau) < t < T \qquad (5.18)$$

Rearranging the time axes to make

$$f_m(t) = 1 \text{ for } -T/2 + \tau/2 < t < T/2 - \tau/2 \qquad (5.19)$$
$$= 0 \text{ for } -T/2 < t < -T/2 + \tau/2, T/2 - \tau/2 < t < T/2 \quad (5.20)$$

$$F\{f_m^*(t)\} = \frac{A}{T}\int_{-\frac{T}{2}+\frac{\tau}{2}}^{\frac{T}{2}-\frac{\tau}{2}} \exp(jn\omega_m t)dt = \frac{A}{T}\int_{-\frac{T'}{2}}^{\frac{T'}{2}} \exp(jn\omega_m t)dt \qquad (5.21)$$

where n = harmonic number

$$F\{f_m^*(t)\} = \frac{AT'}{T}\frac{\sin\{(n\omega_m T')/2\}}{(n\omega_m T')/2} \qquad (5.22)$$

where:

$$T' = (T - \tau)$$
$$F\{f_m^*(t)\} = A(T - \tau)\frac{\sin\{n\omega_m(T - \tau)/2\}}{n\{\omega_m(T - \tau)/2\}} \qquad (5.23)$$

Hence the IF spectrum is given by:

$$f_{IF} = A\left\{-j\pi\delta\left(\omega - \omega_{if}\right) + j\pi\delta\left(\omega + \omega_{if}\right)\right\}^* A\frac{(T - \tau)}{T}\frac{\sin n\omega_m\{(T - \tau)/2\}}{n\omega_m(T - \tau)/2}$$
$$(5.24)$$

It can be seen that the periodic IF signal consists of an envelope function $(\sin x/x)$ enclosing a line function. In essence, the FMCW radar measures the phase of the IF signal which is directly related to target range. The frequency of

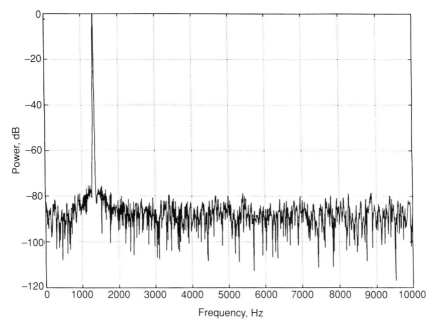

Figure 5.18   IF power spectrum for zero sweep nonlinearity (courtesy ERA Technology)

the IF signal can be regarded as a measure of range. An inverse frequency–time transform can reproduce a time-domain equivalent to the impulse radar.

The FMCW radar system is particularly sensitive to certain parameters. In particular it requires a high degree of linearity of frequency sweep with time to avoid spectral widening of the IF and hence degradation of system resolution.

This effect can be illustrated by considering the case of a FMCW radar with varying degrees of nonlinearity as shown in Figures 5.18 and 5.19. The IF spectrum is shown for two cases: (a) perfect (zero) nonlinearity, and (b) 0.5% nonlinearity. The main effect is to broaden the width of the IF spectrum as the extent of nonlinearity increases and increase the sidelobe level. In practice a useful system should aim to keep all nonlinearities less than 0.1%. A further parameter which must be carefully controlled is the purity of the spectral output. Both the phase noise spectrum and inband harmonics should be reduced to low levels. With inband harmonics, the effect is to generate clutter while for phase noise the effect is to reduce the sensitivity of the radar for adjacent targets.

The dynamic performance of an FMCW radar for a triangular sweep waveform can be estimated from the example given in Table 5.2.

A plot of received signal against range is shown in Figure 5.20. From this it can be seen that the operating range of the radar system lies between the boundaries defined by the system-dynamic-range clutter profile and target reflection loss. The greater dynamic range of the FMCW radar is a significant advantage provided that the sweep linearity can be maintained and the spectral

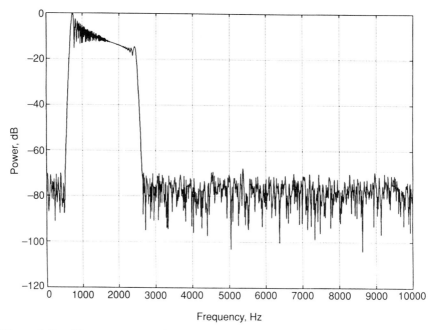

Figure 5.19   IF power spectrum for 0.5% sweep nonlinearity (courtesy ERA Technology)

broadening and sidelobes of the IF envelope minimised. A plot of the ratio of peak to sidelobe level as a function of linearity for several ranges (Dennis and Gibbs, 1974) is shown in Figure 5.21.

It is also important that the frequency of the output is stable over time as instability reduces the calibration accuracy and hence comparison of measurements taken at different times cannot be carried out.

There are other particular features of an FMCW radar system which must be

Table 5.2   Characteristics for an FMCW radar

Transmitter:	Mean power	= 500 mW (26.9 dBm)
	Antenna and cable losses	= −9 dB
	Mean radiated power	= 61 mW (17.9 dBm)
	Repetition rate	= 1 ms
	Clutter profile	= 10 dB/ns
Receiver:	RF bandwidth	= 1 GHz
	IF bandwidth	= 13.3 kHz per metre (free space)
	Equivalent thermal noise	= 5.52×10^{-17} per metre (−132 dBm)
	Mixer noise figure	= 8 dB
	Minimum signal level	= −124 dBm
	Maximum signal level	= +10 dBm
	Dynamic range	= 134 dB

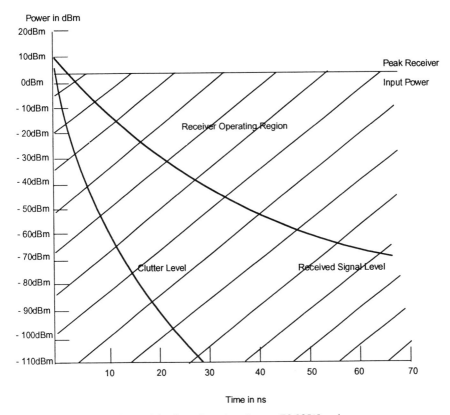

Figure 5.20   Received signal level against time for an FMCW radar

considered. With reference to Figure 5.15, the system block diagram, it will be seen that a buffer amplifier is incorporated. The purpose of this is to reduce the effect of changes in VSWR over the range of swept frequencies causing 'frequency pulling' of the VCO. Changes in VSWR can be caused by variations in antenna-to-surface spacing or by changes in the characteristics of components such as the circulator or mixer. The isolation (S_{12}) of the buffer amplifier should be sufficient to reduce frequency pulling to insignificant levels.

The amplitude–frequency transfer characteristics of all of the components in an FMCW radar system should be substantially flat. Ideally, amplitude-ripple levels should be less than ± 0.25 dB, otherwise the radiated waveform will exhibit an amplitude modulation which will cause spectral spreading of the IF waveform with a resulting loss of solution and system performance.

The FMCW radar shown in Figure 5.15 contains a number of features to enable the linearity of the radar system to be optimised. The first element is a microwave discriminator which provides an output voltage whose frequency should be a constant. Variation from a fixed value provides an error voltage which is used to compensate the drive voltage for the microwave voltage-

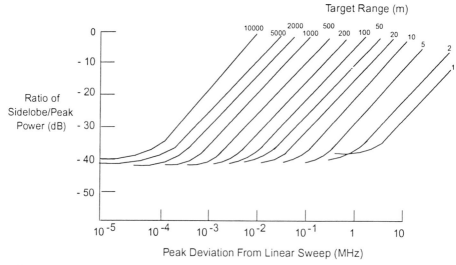

Figure 5.21   *Peak-to-sidelobe ratio as a function of sweep nonlinearity (after Dennis and Gibbs)*

controlled oscillator. Provided that the delay time in the discriminator is short, the loop bandwidth of the control loop is adequate and the rate of sweep is appropriately selected, the sweep waveform can be compensated in real time for the inevitable nonlinearities which are found with a varactor-diode tuning element. Such devices do not exhibit a linear capacitance/tuning voltage law and require either static or dynamic linearisation.

In the block diagram shown, a tracking filter is included to acquire information from the air–material interface reflection and use this to calibrate the radar.

The output from the IF will, for the antenna transmitting into an infinite lossy dielectric halfspace, contain signals due to imperfections in the radar system itself. These will be chiefly caused by, in order of significance, reflected energy from the antenna mismatch, leakage from port 1 to port 3 of the circulator and leakage via the directivity of the coupler.

As most of these signals will be at very low frequencies, a highpass filter in the IF signal path can be used to reduce their effect. The same filter can be used to compensate for the spreading loss in free space encountered by the radiated and reflected signal. In voltage terms, an R^{-2} variation translates to a highpass filter with 12 dB-per-octave attenuation characteristics.

For a complex mixer with I/Q outputs, similarity of the mixer characteristics is important and a well matched pair should be selected.

The complex output from the mixer consists of a set of frequencies representing reflections from individual targets. If an inverse Fourier transform is carried out, a time-domain representation of the target reflections can be generated. However, it is well known that the resolution of the fast Fourier transform is suboptimum for most spectral analysis and a range of alternative

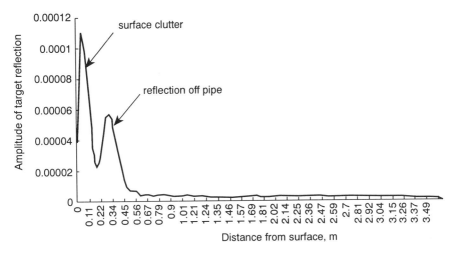

Figure 5.22 Typical FMCW-radar received signal against time (courtesy ERA Technology)

transforms can be used to obtain improved resolution. In a classic paper Kay and Marple (1981) discussed alternative transform techniques and showed that the following methods could provide improved performance compared with the FFT. This aspect is discussed in greater detail in Chapter 7 on signal processing.

Typical examples of frequency-modulated signals received from buried targets are shown in Figures 5.22 and 5.23. Further references to FMCW radars are found in papers by Adler and Jacobs, 1993; Al-attar *et al.*, 1982; Botros *et al.*, 1984; Carr *et al.*, 1986; Farmer *et al.*, 1984; Garvin and Inggs, 1991; Hua *et al.*, 1991; Ji-Chang, 1990; Olver *et al.*, 1982; Olver and Cuthbert, 1988; Peebles and Stevens, 1965; Stove, 1992; Transbarger, 1985; Yamaguchi *et al.*, 1990; 1991; 1992.

5.4 Synthesised or stepped-frequency radar

A synthesised system is essentially a stepped-frequency continuous-wave radar (Inggs and Garriss, 1992; Ilzuka and Freundorfer, 1984). Any repetitive pulsed signal can be transformed to a frequency-domain representation which will consist of a line spectrum whose frequency spacing is related to the pulse-repetition rate and envelope is related to the pulse shape. Hence a repetitive impulsive waveform can be synthesised by transmitting a sequential series of individual frequencies whose amplitude and phase are accurately known.

Two forms of the synthesised radar can be considered. The first and simplest system is a stepped-frequency continuous wave as shown in Figure 5.24. The

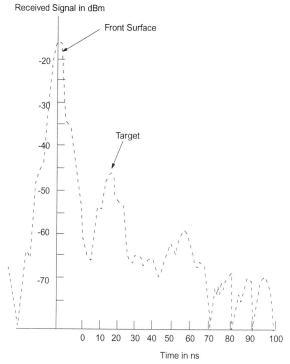

Figure 5.23   Typical FMCW-radar received signal against time (log-amplitude scale)

second form is more complex in that each individual frequency is appropriately weighted in amplitude and phase prior to transmission.

In both cases the difference frequency is of course composed of contributions from all targets up to and beyond the ambiguous range given by

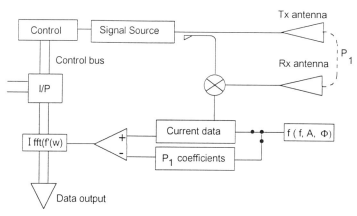

Figure 5.24   Block diagram of a stepped-frequency radar system

$$R_{amb} = \Delta R N \qquad (5.25)$$

where

$\Delta R =$ range resolution

$N =$ number of frequency steps

The radar radiates a sequence of N frequencies and the amplitude and phase of the received and downconverted signal are stored. A complex inverse fast Fourier transform or equivalent algorithm is then used to produce a time-domain version of the reflected signal.

The range resolution is given by

$$\Delta R = \frac{c}{2N\Delta f} \qquad (5.26)$$

For example, a radar using 256 steps with frequency increment of 5 MHz would achieve a range resolution in air of

$$\Delta R = \frac{3 \times 10^8}{2 \times 256 \times 5 \times 10^6} = 0.117 \text{ m}$$

and an ambiguous range

$$R_{amb} = 29.95 \text{ m}$$

If the signal transmitted by the radar is given by

$$E'_T = E_T \, e^{-j\omega t} \qquad (5.27)$$

for a lossy medium and a single target the received signal can be considered as

$$E_{r_0} = \frac{E_T}{r_0^2} \sigma \exp\{j(2kr_0 - \omega t)\} \qquad (5.28)$$

where

$r_0 =$ range to the target
$\sigma =$ target scattering cross-section
$k =$ propagation coefficient

The received signal can be represented by a phasor, as shown in Figure 5.25. As

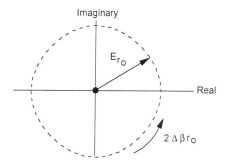

Figure 5.25   *Phasor representation of the measured signal from a single target*
If transmitted frequency changes, then β changes, hence E_{r0} rotates
r   = range
σ   = target cross-sectional area
β   = propagation constant
E_T = transmitted field strength

$$E_{r_0} = \frac{E_T}{r_0^2} \, \sigma \exp\{j(2\beta r_0 - \omega t)\}$$

the operating frequency is changed such that k changes by Δk, the phasor rotates by an angle equal to $2\Delta k r_0$.

Evidently, the amount of rotation is related to the target range. If the frequency of the transmitted signal is incrementally increased, the frequency of the received signal will be related to the range, taking into account the propagation coefficient.

For multiple targets the received signal becomes the phasor sum of all contributions, as shown in Figure 5.26:

$$E_r' = \sum_{n=0}^{N-1} E_{rn} \tag{5.29}$$

The range to each target can be determined by performing a suitable transform with respect to the steps of frequency. The received signal can be expressed as

$$E_n = \sum_{k=0}^{N-1} \frac{\sigma_k E_T}{r_k^2} \exp j4\pi\{(f_0 + n\Delta f)r_k\}/c \tag{5.30}$$

where

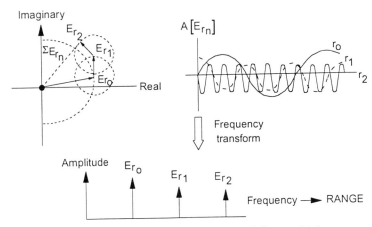

Figure 5.26 Phasor representation of the received signal from multiple targets

$$
\begin{aligned}
r_k &= \text{distance to the } k\text{th target} \\
\sigma_k &= \text{target scattering coefficient} \\
f_0 &= \text{start frequency} \\
\Delta f &= \text{frequency increment} \\
n &= \text{number of frequency steps}
\end{aligned}
$$

This equation can be seen to take the form of a discrete Fourier transform and therefore the scattering magnitude of σ_k can be obtained using a fast Fourier transform to indicate depth, in quantised steps, of a target.

A synthesised system may be operated such that the received signal may be integrated for as long as convenient in relation to the radar survey speed. Typically, each frequency might be transmitted for a time of 100 μs and the resulting intermediate frequency integrated for this interval, thus providing a receiver bandwidth of

$$B = \frac{1}{\tau} \tag{5.31}$$

which for this time would equal to 10 kHz.

It is evident that the receiver thermal noise over this bandwidth is very much lower than that of the receiver of the time-domain radar and an improvement of typically 50 dB in noise performance compared with the latter can be achieved.

For a synthesised radar, contributions to the received signal outside the ambiguous range cannot be removed by simple anti-aliasing filters.

It is therefore important to choose operating parameters which minimise the aliasing effect. One method is to determine the range gate of the target iteratively. The initial measurement is taken with low resolution and large ambiguous range. The range to the target is determined and the resolution is then increased.

Normally the radar is calibrated both to establish a reference plane for measurement as well as to reduce the effect of variation of the frequency characteristics in components and antennas.

The radar system will introduce additional phase shifts on the transmitted and received systems. These will be caused by the electrical lengths of the signal paths to the antennas and the effective radiation phase centre of the antennas. This means that the phases of the transmitted and received signals will be different at each integer frequency and will require compensation.

One method of calibrating the radar (Ilzuka and Freundorfer, 1984) is to place a known reflector at a defined distance from the antennas. A metallic plane is suitable. The compensation factors can be adjusted such that the indicated range then equates to the actual range. Essentially,

$$E_{ref} = \frac{E_T}{r_{ref}^2}\sigma\exp\{j(2kr_{ref} - \omega t) + j\phi_{ref}\} \tag{5.32}$$

It is evident that it is important that the recorded values of amplitude and phase at each frequency are accurately related. Any temporal variation in the system characteristics which degrades the system calibration will, of course, reduce the resolution and accuracy of measurement.

In addition, the repeatability of the frequency is important and must be such that the calibration remains valid.

The main advantages of a stepped-frequency continuous-wave radar are: its ability to adjust the range of frequencies of operation to suit the material and targets under investigation, a higher mean-radiated-power level per spectral line and the ability to integrate the received signal level, hence improving the system sensitivity.

The calibration of the radar does, of course, depend on stable system characteristics and antenna parameters that are invariant with front surface-antenna spacing.

The synthesised pulse radar is a variant of the step-frequency continuous-wave radar in that the relative amplitudes and phases of the transmitted frequencies are adjusted on transmission in order to synthesise the desired pulse waveform.

A significant advantage for this approach is that it is possible, within limits, to take account of the frequency characteristics of the antennas. For example, the low-frequency cutoff point of an antenna can be compensated by weighting the amplitude of the low-frequency spectral lines.

However, the requirement to maintain an accurate phase relation between each of the spectral lines is difficult to achieve in real time. It is generally easier to carry out computation on downconverted and recorded data; hence the stepped-frequency continuous-wave radar is a more economic design option.

5.5 Single-frequency methods

Single-frequency methods of imaging are based on the technique of viewing the target from a number of physically different positions in an aperture over the target, recording the amplitude and phase of the received signal and then mathematically reconstructing an image of the radiating source. Essentially, a synthetic aperture is constructed over a defined aperture at the measurement plane.

In general, most antennas used in surface-penetrating radar have relatively low gain and hence a wide beamwidth which results in poor resolution in either x or y dimensions. Synthetic-aperture methods aim to increase x or y resolution by synthetically generating an antenna with a large aperture and consequent reduced beamwidth.

The synthetic-aperture array must consist of a minimum number of samples in order to avoid aliasing. The minimum number is given (Osumi and Ueno, 1984) by

$$N = \frac{2R\beta^2}{\lambda} \tag{5.33}$$

where:

$$\begin{aligned}
\beta &= \text{antenna beamwidth in radians} \\
R &= \text{range to the target} \\
\lambda &= \text{radiated wavelength}
\end{aligned}$$

There are several variations on the synthetic-aperture method which can be considered. Where a single frequency is used, holographic methods can be employed to generate an image. The holographic method records the amplitude and phase of the received signal in a plane over the target. This function is then correlated with a test function which is set to provide a maximum value of the crosscorrelation where reflection occurs and to be zero elsewhere.

The holographic image-reconstruction method is defined in the two-dimensional case by the correlation between the test function $h(x, t)$ and received signal $v_r(x_r \cdot t)$ where x is the co-ordinate vector of the imaging point and x_r is the co-ordinate vector of the receiving point, as shown in Figure 5.27 and $v_r(x_r t)$ is the signal received at the receiving point.

The linear operation to reconstruct the image is defined (Osumi and Ueno, 1984) by

$$b(x) = \int \int \int_{-\infty}^{\infty} v_r(x_r, t) h(x - x_r, t) dt \; dx_r \; dy_r \tag{5.34}$$

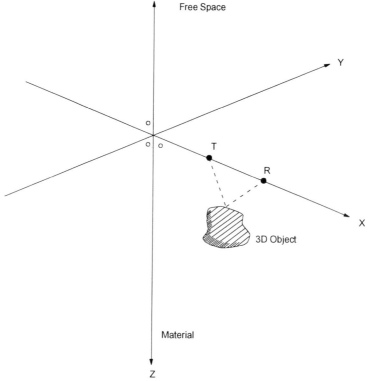

Figure 5.27   Co-ordinates of holographic imaging systems

as the received signal can be considered to be

$$v_1(x_r, t) = \frac{1}{2\pi} \int_{-\infty}^{\infty} V(x_r, \omega) \exp j(\omega t) \, d\omega \qquad (5.35)$$

$$v_r(x_r, t) = \frac{1}{2\pi c} \iint_s \frac{\cos \theta}{r^2} \sigma(x) \frac{d}{dt} \frac{(t - 2r)}{c} \, dx dy \qquad (5.36)$$

where $\sigma(x)$ is the reflectivity of the target.
   Hence the image function of the object T

$$b(x) = \frac{1}{2\pi C} \iint_s \sigma(x_r) b_0(x - x_r) dx_r dy_r \qquad (5.37)$$

where

$$b_0(x) = \iiint \frac{\cos\theta}{|x - x_r|^2} \frac{d}{dt} \frac{[t - 2|x - x_r|]}{c} h(x_r, t) dt dx_r dy_r \qquad (5.38)$$

The effect of material attenuation is significant as the general effect is to apply a windowing function across the recording aperture, thus limiting its useful size in relation to sharply focused images, as shown in Figure 5.28.

In addition, the effect of both material attenuation and relative permittivity on the antenna beamwidth should be considered. As the values of loss and relative permittivity increase, the beamwidth of the antenna reduces and this degrades the gain of the synthetic aperture. In general, synthetic-aperture methods are most useful in lower-loss materials. Typical images are shown in Figure 5.29.

Single-frequency methods require accurate recording of $v_r(x_r, t)$ over a line and over the complete aperture. This accurate registration of the data is important and is somewhat difficult to achieve under field conditions. Most previous work (Anderson, 1977) and Osumi and Ueno, 1984) used accurate mechanical scanning X-Y frames and the practical difficulties of using these in real-life conditions have limited the use of single-frequency holographic methods. However, advances in low-cost robotic technology may overcome some of the past hurdles.

Alternative methods of multifrequency image reconstruction are based on diffraction stack migration, Kirchoff's methods and Green's-function methods which will be referred to in Chapter 6 on signal processing.

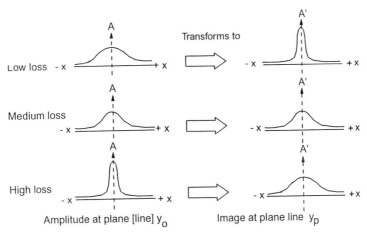

Figure 5.28 Effect of material attenuation on SPR synthetic-aperture image

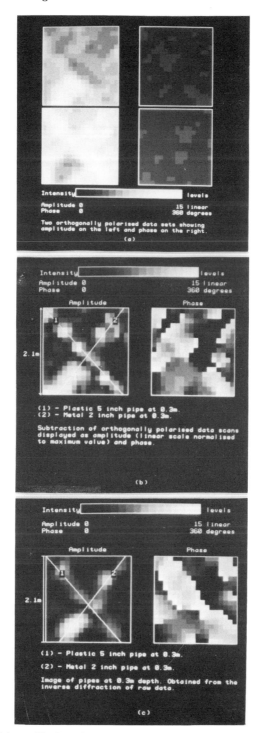

Figure 5.29   *Holographic data from two separate orthogonal scans (a) is subtracted (b)
and focused to the object plane (c) (Junkin)*

5.6 Polarisation modulation

Any target which possesses a linear scattering geometry can be identified by means of its polarisation scattering matrix. For a linear target

$$\mathbf{E}^s = \mathbf{S}\mathbf{E}^i \qquad (5.39)$$

where

$$\mathbf{E}^s = \begin{bmatrix} E_x^s \\ E_y^s \end{bmatrix} = \begin{bmatrix} S_{xx} S_{xy} \\ S_{yx} S_{yy} \end{bmatrix} \begin{bmatrix} E_x^i \\ E_y^i \end{bmatrix} \qquad (5.40)$$

and the predominantly linear feature lies in the x axis and is lower in impedance than the surrounding material then the parameter S_{xx} predominates. This is usually the case for buried metal targets, whereas for plastics where the dielectric surround has a higher impedance than the target $S_{yy} > S_{xx}$. As S_{xy} and S_{yx} tend to zero, then the backscattered E field is largely parallel to the target axis in the case of a parallel incident E field applied to a linear metallic target.

When a plastic pipe is illuminated with an orthogonal incident E field, the backscattered E field is the largest component of the reflected signal.

Many systems have exploited this polarisation sensitivity by using orthogonal transmit and receive antennas. The crossed dipole exhibits a much lower crosscoupling than a copolarised pair and this improves the system detectivity.

It is well known that targets such as pipes, as well as shells of various calibres and cracks, act as depolarising features. A linearly polarised crossed-dipole antenna rotated about an axis normal to the target produces a sinusoidal variation in received signal. However, the null points are a distinct disadvantage because the operator is required to make two separate, axially rotated measurements at every point to be sure of detecting pipes at unknown orientations.

If such a crossed dipole is rotated around its own axis, the amplitude of the received signal will vary sinusoidally with the angular rotation of the antenna. Following Daniels *et al.* (1988),

$$E_r = E_t(-\sin\theta \cos\theta) \begin{bmatrix} S_{xx} S_{xy} \\ S_{yx} S_{yy} \end{bmatrix} \begin{bmatrix} \cos\theta \\ \sin\theta \end{bmatrix} \qquad (5.41)$$

Hence if S_{xy} and S_{yx} are neglected then

$$E_r = E_t \left\{ \frac{1}{2} (S_{yy} - S_{xx}) \sin 2\theta \right\} \qquad (5.42)$$

In real life, additional received signals caused by a variety of factors such as changes in crosscoupling between the crossed dipoles due to objects or variations in local impedance on the ground surface will contribute to E_r which can be rewritten as

$$E_r = K_1 \cos\theta + K_2 E_t \left(S_{yy} - S_{xx}\right) \sin 2\theta \qquad (5.43)$$

The need for rotation of the antenna is physically restrictive and electronic means of rotation have been considered.

One design possibility is to synthesise a circularly polarised signal. Any wave of arbitrary polarisation can be synthesised from two waves orthogonally polarised to each other. As shown in Chapter 4, in Figure 4.35 a circularly polarised wave is produced by exciting vertically and horizontally polarised waves, each having the same amplitude and with a 90° phase difference between them.

The radiating elements are fed, via wideband (preferably decade) 180° and 90° hybrids, to radiate circular polarisation. If right-hand circularly polarised signals are transmitted and received, the preferential detection of linear features is achieved. If, however, right-hand circularly polarised signals are transmitted and left-hand circularly polarised signals are received, planar features are detected. Hence, if connections to the radiating elements are arranged and switched appropriately, the signals routed to the receiver contain different data according to the sense of polarisation. The date, therefore, can be processed separately and in a different manner in order to provide images of different targets in the material under investigation.

Unfortunately, hardware deficiencies limit the performance; first, it is difficult to achieve wideband operation with 90° hybrids (at least over a decade) and, secondly, even the fastest state-of-the-art GaAs switches have unacceptably high breakthrough levels.

In view of these difficulties the feasibility of using a commutated multi-element crossed-dipole array can be considered and this technique is rotating linear rather than circular polarisation.

A simple option is an eight-element antenna in which the crossed-dipole pairs can be switched at intervals of up to 1 ms so that the two crossed-dipole pairs are orientated between 0° and 45°, as shown in Figure 4.36. This antenna proved to be a successful demonstration of the proof of the concept.

Full commutation over 360° in 45° steps could be achieved as shown in Figure 4.37 by using PIN-diode switches (to handle the transmitted power) operating at a switching interval of 1 μs, thus achieving 360° rotation in approximately 10 μs. The possibility of real-time discrimination using filters based on recognition of the $\cos 2\theta$ amplitude variation of each range sample is also possible. Operationally, such a system would have the advantage of being able to survey rapidly without the limitations imposed by mechanically rotated antennas.

Conventionally, circular polarisation refers to a steady-state condition during which a long-duration pulse or CW waveforms are transmitted. For impulse radars, the pulse duration is very short (< 5 ns) and hence a more complex transient situation is encountered. In general, several cycles of transmitted wave are needed to establish the state of circular polarisation.

One method of radiating circular polarisation with an impulse waveform is to use an equi-angular spiral antenna. Unfortunately, the dispersive nature of this type of antenna causes an increase in the duration of the transmitted waveforms and the radiated pulse takes the form of a 'chirp' in which high frequencies are radiated first, followed by the low frequencies. This effect, however, may be compensated by a spiking' filter, which may take the form of a conventional matched filter or a more sophisticated filter such as Wiener filter. The use of such an antenna has been shown (Shaw *et al*, 1993) to be a useful method of implementing a pipe-detection radar and plastics pipes buried in wet clay have been detected up to a depth of 1.0 m. Further references are found in Ueno and Osumi (1984); Junkin and Anderson (1986, 1988); Orme and Anderson (1973); Richards and Anderson (1978); Anderson and Richards (1977); Osumi and Ueno (1985, 1988); and Tanaka *et al*. (1985).

5.7 Summary

The vast majority of surface-penetrating-radar systems that have been built are based on the time-domain amplitude-modulated system approach. The general simplicity of concept and relatively lower cost of production when compared with frequency-modulated or step-frequency radar systems have been powerful reasons until now for the choice of impulse-radar methods. However, the technical performance of impulse-radar systems is generally severely limited by the receiver which is a sampling downconvertor exhibiting a poor noise figure and conversion efficiency.

For this reason attention has been paid to alternative designs. The continuous-wave frequency-modulated system design requires a high performance in terms of linearity of frequency sweep with time and this is often difficult to achieve within tight budget constraints. The step-frequency continuous-wave radar offers considerable promise for the future now that the performance of synthesised frequency sources has improved and their cost has reduced. The major advantage of the frequency-modulated systems is their improved receiver performance in terms of dynamic range compared with the impulse-radar system.

Single-frequency systems using holographic image-reconstruction techniques have been shown to be viable but the physical difficulty of accurately recording over an aperture has tended to limit their use.

140 *Surface-penetrating radar*

5.8 References

ADLER, D., and JACOBS, M. (1993): Application of a narrowband FM–CW system in the measurement of ice thickness. Proceedings of IEEE Measurement & Testing Society *International Microwave Symposium*, Atlanta, GA, USA, Vol. 2, 809–812

AL-ATTAR, A., DANIELS, D. J., and SCOTT, H. F. (1982): A novel method of suppressing clutter in very short range radars. Proceedings of the IEE *International Radar Conference*, RADAR–82, London, UK, 419–423

ANDERSON, A. P. (1977): Microwave holography. *Proc. IEE*, **124**, (11R), 946–962

ANDERSON, A. P., and RICHARDS, P. J. (1977): Microwave imaging of subsurface cylindrical scatterers from cross–polar backscatter. *Electron. Lett.*, **13**, 617–619

BOTROS, A. Z., OLVER, A. D., CUTHBERT, L. G., and FARMER, G. (1984): Microwave detection of hidden objects in walls. *Electron. Lett.*, **20**, 379-380

CARR, A. G., CUTHBERT, L. G., and LIAU, T. F. (1986): Signal processing techniques for short–range radars applied to the detection of hidden objects. Proceedings of 7th European conference on *Electrotechnics*, EURCON '88, Paris, France, 641–646

DANIELS, D. J., GUNTON, D. J., and SCOTT, H. F. (1988): Introduction to subsurface radar. *IEE Proc. F*, **135**, (4), 278–321

DENNIS, P., and GIBBS, S. E. (1974): Solid-state linear FM/CW radar systems - their promise and their problems. Proceedings of *IEEE International Microwave Symposium*, Atlanta, GA, USA, 340–342

FARMER, G. A., CUTHBERT, L. G., OLVER, A. D., BOTROS, A. Z. (1984): Distinguishing between types of hidden objects using an FMCW radar. *Electron. Lett.*, **20**, 824–825

GARVIN, A. D. M., and INGGS, M. R. (1991): Use of synthetic aperture and stepped frequency continuous wave processing to obtain radar images. Proceedings of South African symposium on *Communications and Signal Processing*, COMSIE '91, Pretoria, S. Africa, 32–35

HARES, J. D. (1992): Applications for solid-state high-voltage pulsers. *Eng. Sci. Educ. J.*, June, 113–120

HUA, L., COOPER, D. C., and SHARMAN, E. D. R. (1991): A prototype FMCW radar using an analogue linear frequency sweep. Proceedings of IEE colloquium on *High Time-Bandwidth Product Waveforms in Radar and Sonar, IEE Colloquium Digest 093*, 11/1–11/3

ILZUKA, Z., and FREUNDORFER, A. (1984): Step frequency radar. *J. Appl. Phys.*, **56**, 2572–2583

INGGS, M., and GARRISS, A. (1992): A stepped frequency CW ground probing radar. MRSI Microwave signature 92, 1–4

JI-CHANG, F. (1990): A new method of signal processing with high range resolution and reduced wide–bandwidth for linear–FMCW radar. *AMSE Rev.*, **14**, 59-63

JUNKIN, G., and ANDERSON, A. P. (1986): A new system for microwave holographic imaging of buried services. Proceedings of 16th European Microwave Conference, 720–725

JUNKIN, G., and ANDERSON, A. P. (1988): Limitations in microwave holographic synthetic aperture imaging over a lossy half-space. *IEE Proc. F*, **135**, 321–329

KAY, S., and MARPLE, S. L. (1981): Spectrum analysis. A modern perspective. *Proc. IEEE*, **69**, (11), 1380–1419

LEE, C. H. (1991): Generation of high power ultrawideband electrical impulse by optoelectronic technique. Proceedings of IEEE Measurement & Testing Society *International Microwave Symposium*, Boston, MA, USA, Vol. 1, 375–378

OLVER, A. D., CUTHBERT, L. G., NICOLAIDES, M., and CARR, A. G. (1982): Portable FMCW radar for locating buried pipes. Proceedings of *Radar '82* conference, London, UK, *IEE Conf. Publ. 216*, 413–418

OLVER, A. D., and CUTHBERT, L. G. (1988): FMCW radar for hidden object detection. *IEE Proc. F*, **135**, 354–361

ORME, R. D., and ANDERSON, A. P. (1973): High-resolution microwave holographic technique: application to the imaging of objects obscured by dielectric media. *Proc. IEE*, **120**, 401–406

OSUMI, N., and UENO, K. (1984): Microwave holographic imaging method with improved resolution. *IEEE Trans.*, **AP–32**, 1018–1026

OSUMI, N., and UENO, K. (1985): Microwave holographic imaging of underground objects. *IEEE Trans.*, **AP-33**, 152–159

OSUMI, N., and UENO, K. (1988): Detection of buried plant. *IEE Proc. F*, **135**, 330–342

PEEBLES, P. Z., and STEVENS, G. H. (1965): A technique for the generation of highly linear FM pulse radar signals. *IEEE Trans.*, **MIL-9**, 32–38

POCHA, M. D., DRUCE, R. L., WILSON, M. J., and HOFER, W. W. (1989): Avalanche photoconductive switching. Proceedings of 7th IEEE *Pulsed Power Conference*, Monterey, CA, USA, 866–868

RICHARDS, P. J., and ANDERSON, A. P. (1973): Microwave images of sub-surface utilities in an urban environment. Proceedings of 8th *European Microwave Conference*, 33-37

SHAW, M. R., MILLARD, S. G., HOULDEN, M. A., AUSTIN, B. A., and BUNGEY, J. H. (1993): A large diameter transmission line for the measurement of the relative permittivity of construction materials. *Br. J. Non. Destr. Test.*, **35**, (12), 696–704

STOVE, A. G. (1992): Linear FMCW radar techniques. *IEE Proc. F*, **139**, (5), 343–351

TANAKA, H., KAWONO, A., KOYANAGI, M., and MATSUURA, M. (1985): Development of radar-based pipe locator. Proceedings of 16th *World Gas Conference*, Munich, Germany, International Gas Union

TRANSBARGER, O. (1985): FM radar for inspecting brick and concrete tunnels. *Mater. Eval.*, **43**, (10), 1254–1261

UENO, K., and OSUMI, N. (1984): Underground pipe detection based on microwave polarisation effect. Proceedings of international symposium on *Noise & Clutter Rejection*, Tokyo, Japan, 673–678

YAMAGUCHI, Y., SENGOKU, N., and ABE, T. (1990): FM-CW radar applied to the detection of buried objects in snowpack. Proceedings of IEEE international symposium on *Antennas and Propagation - Merging Technologies for 90s*, Dallas, TX, USA, Vol. 2, 738–741

YAMAGUCHI, Y., MARUYANA, Y., KAWAKAMI, A., SENGOKO, M., and ABE. T. (1991): Detection of objects buried in wet snowpack by an FM–CW radar. IEEE *Trans., Geosci. Remote Sens.*, **29**, (2), 201–208

YAMAGUCHI, Y., MITSUMOTO, M., SENGOKU, M., and ABE, T. (1992): Synthetic aperture FM–CW radar applied to the detection of objects buried in snowpack. Digest of IEEE international symposium on *Antennas and Propagation*, 1122–1125

Chapter 6

Signal processing

6.1 Introduction

The objective of this Chapter is to provide an introduction to those methods that
have been used to process data. The area of signal processing is so extensive that
only a basic introduction to the topic is possible. The recommended references
will provide further material for those wishing to investigate the topic in greater
detail.

Inevitably, some users may consider that their favourite method has not been
given sufficient prominence or that some combinations of methods may provide
improved performance. Unfortunately it is impossible to satisfy all interests.

The author's view is that signal processing is primarily a means of reducing
clutter. The cost–benefit of implementation should be clearly demonstrated
before superficially attractive but practically unsound methods are incorporated. Clearly the wide range of targets, applications and situations encountered
is likely to task even the most robust algorithm and the user should assess the
latest algorithm with some care.

The general objective of signal processing as applied to surface-penetrating
radar is either to present an image that can readily be interpreted by the
operator or to classify the target return with respect to a known test procedure or
template.

The image of a buried target generated by a subsurface radar will not, of
course, correspond to its geometrical representation. The fundamental reasons
for this are related to the ratio of the wavelength of the radiation and the
physical dimensions of the target. In most cases for surface-penetrating radar the
ratio is close to unity. This compares very greatly with an optical image which is
obtained with wavelengths such that the ratio is considerably greater than unity.

In surface-penetrating-radar applications the effect of combinations of
scattering planes, for example, the corner reflector, can cause 'bright spots' in
the image and variations in the velocity of propagation can cause dilation of the
aspect ratio of the image. While many images can be focused to reduce the effect
of antenna beam spreading, regeneration of a geometric model is a much more
complex procedure and is not usually attempted.

An alternative approach which seeks to reduce the workload on the operator
is to correlate the image with a known template and derive a spatial-correlation
coefficient.

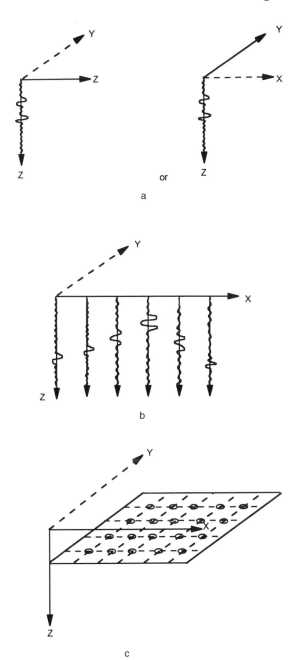

Figure 6.1 Co-ordinate system for scan description
  a A-scan
  b B-scan
  c C-scan

The general procedure when processing data is to store the data in the appropriate dimensional format and then apply appropriate algorithms.

With reference to Figure 6.1 the data can be considered to be of the form

$$A(x_i, y_j, z_k) \tag{6.1}$$

over the ranges

$$k = 1 \text{ to } N$$
$$j = 1 \text{ to } M$$
$$i = 1 \text{ to } P$$

Note that time and depth of the z axis can be considered to be inter-related by the velocity of propagation.

A single waveform or A-scan is defined as

$$\left. f(z) = A(x_i, y_l, z_k) \text{ over the range} \quad \begin{aligned} k &= 1 \text{ to } N \\ i &= \text{ constant} \\ l &= \text{ constant} \end{aligned} \right\} \tag{6.2}$$

An ensemble waveform set or B-scan is defined as

$$\left. \begin{aligned} f(x, z) = A(x_i, y_l, z_k) \text{ over the range } \quad & k = 1 \text{ to } N \\ & i = 1 \text{ to } P \\ & l = \text{ constant} \\ \\ f(y, z) = A(x_i, y_l, z_k) \text{ over the range } \quad & k = 1 \text{ to } N \\ & l = 1 \text{ to } L \\ & i = \text{ constant} \end{aligned} \right\} \tag{6.3}$$

or

An ensemble waveform set or C-scan is defined as

$$\left. f(x, y, z) = A(x_i, y_j, z_k) \text{ over the range } \begin{aligned} j &= 1 \text{ to } M \\ i &= 1 \text{ to } P \end{aligned} \right\} \tag{6.4}$$

for a selected value of z

or range of values of z

In general most published surface-penetrating-radar data have been processed and presented in A, B or C-scan form. The processes applied to each data format can be broadly classified as follows:

(*a*) *A*-scan processing
(*b*) *B*-scan processing. Note that the dimensions can be considered interchangeable, i.e. the same processes may be useful for *x–z*, *x–y* or *y–z* planes.
(*c*) *C*-scan processing. Note that the spatial three-dimensional data can be used to reconstruct representations of three-dimensional images.

Most of the description in this Chapter is relevant to a time series data set. Most amplitude-modulated radar systems generate time-domain data but frequency-domain radar systems such as FMCW initially generate a frequency-domain data set which requires transformation to the time-domain equivalent. Other modulation schemes such as pseudorandom coding or noise modulation require a crosscorrelation of the received signal with a template.

There is no fundamental difference in the information content of the data output from any modulation scheme provided, of course, that the amplitude and phase information are retained in the receiver conversion process.

In general it is the range to the target which is of most interest and, as this is fundamentally equivalent to time, the time-series data set is most relevant. However, there will be occasions when it is more appropriate to use alternative descriptions.

Many of the processing techniques that have been applied to surface-penetrating-radar data have been developed for other applications and, in addition to radar, acoustic, ultrasonic and seismic processing methods have been freely employed.

In reality the time series under consideration represents only the sampled

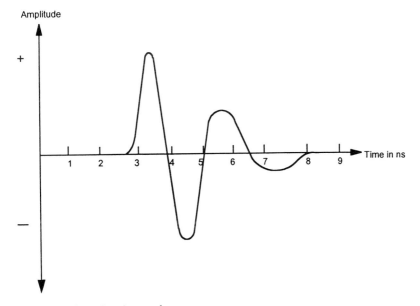

Figure 6.2   Time-domain wavelet

values of a continuous time function and hence is a restricted version of the latter. In addition, it has a limited time duration compared with an analytic function which can be considered to have infinite time duration. The finite number of samples stored and processed represents a truncated portion of a time series and can be termed a sample time series.

A wavelet can be considered as a one-sided transient event with a definite time of arrival and a finite energy content, as shown in Figure 6.2.

The general processing problem encountered in dealing with surface-penetrating-radar data is in the widest sense the extraction of a localised wavelet function from a time series which displays very similar time-domain characteristics to the wavelet. This time series is generated by signals from the ground and other reflecting surfaces, as well as internally from the radar system.

Unlike conventional radar systems in which the target can generally be regarded as being in motion compared with the clutter, in the surface-penetrating-radar case, the target and the clutter are spatially fixed and the radar antenna is moved with respect to the environment.

In all the following discussions it is assumed that data are recorded to an adequate resolution and bandwidth. This means digitisation to at least 12 or 16 bits (72 dB or 96 dB system dynamic range) for an *A*-scan record length of typically 256 or 512 samples. This entails a data-storage capacity of at maximum 1 kbyte per *A*-scan unless data-compression techniques are used.

The capacity required to store a *B*-scan depends upon the overall length of line survey and spatial sampling interval. To achieve an adequate spatial resolution depends on the size of the object and, typically, a convenient length is 512 *A*-scans; hence a *B*-scan may require 0.512 Mbyte of data storage. With the data-storage capabilities of hard disk drives (200–600 Mbyte) and tape back-up stores (1–16 Gbyte) no serious difficulties should be encountered in storing site-survey data. Alternatively, data-compression techniques can be used to provide up to a 20:1 level of data compression.

Before discussing the various signal-processing methods it is useful to state some basic definitions.

The statistical properties of a time series can be determined by considering either a large number of similar signals at any one instant in time, which is termed an ensemble value; or alternatively one signal at a number of intervals of time, termed a time value.

If the two properties are equal then the function or set of functions is said to be ergodic. To be ergodic a function must be a stationary signal, although the converse is not necessarily true.

A continuous function with the same long-term properties is defined as a statistically stationary function and suitable examples are sinusoidal functions or white noise. In contrast, impulsive signals are considered to be statistically nonstationary.

A random signal which has a definite probability as to its content is termed a stochastic signal and, for example, white noise plus a sine wave is termed a stochastic signal.

A minimum-phase system or function is defined as having no poles or zeros in the right-hand half of the *S*-plane. Note that right-hand zeros make a nonminimum phase system but right-hand poles make an unstable system. Conjugate pairs on the *jω* axis indicate a marginally stable system.

6.2 A-scan processing

The received time waveform can be described as the convolution of a number of time functions each representing the impulse response of some component of the radar system in addition to noise contributions from various sources.

Hence the received time waveform (Daniels *et al.*, 1988) is

$$f_r(t) = f_s(t) * f_{a1}(t) * f_c(t) * f_g(t) * f_t(t) * f_g(t) * f_{a2}(t) + n(t) \qquad (6.5)$$

where

$$f_s(t) = \text{signal applied to the antenna}$$
$$f_{an}(t) = \text{antenna impulse response}$$
$$f_c(t) = \text{antenna cross coupling response}$$
$$f_g(t) = \text{ground impulse response}$$
$$f_t(t) = \text{impulse response of target set}$$
$$n(t) = \text{noise}$$

Each contribution has its own particular characteristics which need to be considered carefully before application of a particular processing scheme.

Ideally, the signal applied to the antenna should be a Dirac function but in practice it is more like a skewed Gaussian impulse of defined time duration.

Most antennas used in surface-penetrating applications have a limited low-frequency response and tend to act as highpass filters, effectively differentiating the applied impulse, and hence creating a wavelet. In most cases near-identical antennas are used and if these are spaced sufficiently far from the ground surface then $f_{a1}(t) = f_{a2}(t)$. For antennas operated in close proximity to the ground, both $f_{a1}(t)$ and $f_{a2}(t)$ are variant with changes in the ground-surface electrical parameters.

Any processing scheme which relies on invariant antenna parameters should take into account the mode of operation of the antennas and the degree of stability that is practically realisable.

The antenna crosscoupling response $f_c(t)$ is composed of a fixed contribution $f_c'(t)$ due to antenna crosscoupling or reflection and a variable contribution $f_c''(t)$, the effect of the ground or nearby objects. Hence $f_c(t) = f_c'(t) * f_c''(t)$. It has been found possible to reduce the amplitude of $f_c'(t)$ to very low levels; for crossed-dipole antennas to below −70 dB and for parallel-dipole antennas to below −60 dB. However, $f_c''(t)$ can be significantly larger and degrades the overall value of $f_c(t)$ to −40 dB. The value of $f_c''(t)$ is determined by any local inhomogeneities in the soil or by any covering material whether of mineral or vegetable origin. There is, unfortunately, little that can be done to predict variations in $f_c''(t)$ and it is not amenable to treatment by many processing algorithms. The variation in $f_c''(t)$ is much greater with the crossed-dipole antenna than with the parallel dipole.

The ground impulse response $f_g(t)$ can be determined from its attenuation and relative permittivity across the frequency range of interest.

The target impulse response can be composed of the convolution of the wanted target response, together with many other reflectors which may not be wanted by the user but which are valid reflecting targets as far as electromagnetic waves are concerned. The time separation of the targets is related to their physical spacing, as well as to the velocity of propagation, which can vary depending on the material properties.

Where the targets are well separated in range, it is relatively straightforward to separate the radar reflections but this becomes progressively more difficult as targets become closer together, as instead of individual delta functions a series of overlapping (in time) wavelet functions of different characteristics exists.

Before the more sophisticated methods of recovering and processing the wavelets are considered, there is a range of simple processing methods which can be applied to each individual *A*-scan.

(a) *Zero-offset removal*

An important process operation is to ensure that the mean value of the *A*-scan is near to zero. This assumes that the amplitude probability distribution of the *A*-scan is symmetric about the mean value and not skewed and that the short-time mean value is constant over the time duration of the *A*-scan, as shown in Figure 6.3. It is generally less likely that the amplitude probability distribution is

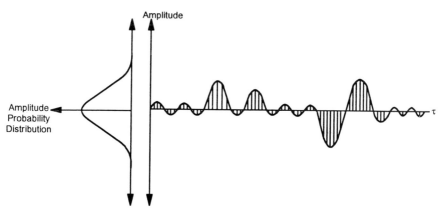

Figure 6.3   A-Scan-sample time series

skewed, but where RF time-varying gain is incorporated and the sampling gate exhibits a DC offset, the short-time mean value may vary over the time duration of the *A*-scan.

Any signal-processing algorithm should cater for these situations. For example, a simple algorithm such as

$$A'_n(t) = A_n(t) - \frac{1}{N}\sum_{n=0}^{N} A_n(t) \tag{6.6}$$

where

$$A_n(t) = \text{unprocessed data sample}$$
$$A'_n(t) = \text{processed data sample}$$
$$\text{and } n = \text{sample number}$$

will only work where the short-term mean value is constant and the amplitude-probability distribution is symmetric.

(b) Noise reduction
An important processing technique is noise reduction and this can be achieved by either averaging each individual sample of the *A*-scan or storing and averaging repeated *A*-scans. The general effect is to reduce the variance of the noise by $\sqrt{N}$ and gives an improvement in signal-to-noise ratio equal to $10\log_{10} N$ dB.

The general form of the filtering operation is given by

$$A'_n(t) = A'_{n-1}(t) + \frac{\{A_n(t) - A'_{n-1}(t)\}}{K} \tag{6.7}$$

where

$$A'_n(t) = \text{averaged value}$$
$$A_n(t) = \text{current value}$$

The factor K may be chosen to be related to n, N or a fixed value which will weight the averaged value appropriately.

The time spent averaging must take into account the speed at which the radar is moved as, unless the physical location is sensibly constant during averaging, data will become contaminated by adjacent samples. Averaging has no effect on clutter but reduces random noise. It is likely that the main contribution to the overall noise level is caused by the radar receiver and this can be very high for the time-domain sampling receiver. In this case much of the noise is caused by timing jitter and can be considered to be worse on the rise and fall times of signals. The noise is, therefore, spectrally biased towards the high-frequency region of the band of frequency operation.

(c) Clutter reduction
Clutter reduction can be achieved by subtracting from each A-scan an averaged value of an ensemble of A-scans or B-scans taken over the area of interest, i.e.

$$A'_{n,a}(t) = A_{n,a}(t) - \frac{1}{N_a} \sum_{a=1}^{N_a} A_{n,a}(t) \qquad (6.8)$$

where

$$n = 1 \ to \ N \ (N = \text{number of samples})$$
$$a = 1 \ to \ N_a \ (N_a = \text{number of } A\text{-scan waveforms})$$
$$A_{n,a}(t) = \text{unprocessed } A\text{-scan}$$
$$A'_{n,a}(t) = \text{processed } A\text{-scan}$$

This method works well for situations where the number of targets is limited and they are physically well separated. It is evident that the summation of the average value will include contributions from all targets, and the greater the number the less will be the difference that results.

In situations where there is a planar interface within the area of interest, this process has the unfortunate effect of removing most of the wavelet caused by the interface. Hence it is important to choose both N and N_a with care to optimise the process for a particular situation.

Fundamentally it is assumed that the material properties vary in a random manner over the volume of interest and the averaged value $f_r(t)$ represents the convolution of

$$f_r(t) = f_s(t) * f_{a1}(t) * f_c(t) * f_g(t) * f_g(t) * f_{a2}(t) \qquad (6.9)$$

The main uncertainty lies in the lack of random variability of the ground properties and the effect this has on $f_r(t)$.

Analysis of the statistical nature of the variability has been carried out by Caldecott *et al.* (1985). This method derives a standard-deviation time function from an ensemble of zero-mean A-scans. Each unprocessed A-scan is compared with the standard-deviation time function and samples which are greater than the standard-deviation time function by a predetermined amount, i.e. one, two or three times σ are defined as significant. An alternative version of this process is

$$A'_{n,a}(t) = A_{n,a}(t) - \frac{K}{N_a} \sum_{a=1}^{N_a} A_{n,a}(t) \qquad (6.10)$$

where K is a variable related to the required magnitude of the standard deviation.

This process can also be applied to a selected section of the A-scan in order to remove clutter associated with a particular region of time. For example, the antenna response can be removed following acquisition of an A-scan from a calibrated target.

(d) Time-varying gain
The received signal is reduced in amplitude compared with the transmitted signal as a result of attenuation both by the medium of propagation and by the path or spreading loss encountered in travelling to and from the target.

In order to apply time-varying gain to compensate for these losses, several conditions must be met. There should be a zero-mean value of the A-scan, otherwise significant DC offsets will be created at late times. The noise levels at late times should be low, else the general late-time noise will be increased. Great care is needed to apply time-varying gain in a smoothly and continuously varying way which corresponds to correction of physical loss mechanisms.

Stepped or rapid gain variation along the time axis can modulate the unprocessed signal and generate artificial wavelets. In general, such variations are to be avoided unless there has been careful assessment of the propagation-path losses as a function of time.

It is more prudent to record data that has a known time-varying gain. This reduces the possibility of data becoming subjectively 'improved' by the field operator of the radar.

(e) Frequency filtering
Highpass filtering of the A-scan data is a useful means of improving the signal-to-clutter ratio in situations where clutter is caused by additional low-frequency energy generated by antenna–ground interactions. In addition, excessive high-frequency noise can usefully be reduced by lowpass filtering. Many commercially available radar systems offer a range of filter options and the choice of settings of bandwidths, slope etc. is left to the operator. Such a filter should in general exhibit a minimum-phase response to reduce phase distortion of the filtered wavelets.

(f) Wavelet optimisation or deconvolution techniques
The general principle of wavelet optimisation is to filter the sample time series in such a way that the desired output from the filtering process is an impulse function representing the deconvolution of the wanted signal. Such an impulse is variously known as a Dirac or delta function, a unit impulse or spike. However, this ideal outcome is not usually achievable and the optimum filter is one where the energy of the difference between the desired and actual filters is minimised. This is termed an optimum or least-squares filter.

$$E^2 = \frac{\lim}{t \to \infty} \frac{1}{2t} \int_{-t}^{t} \{y(t) - d(t)\}^2 dt \qquad (6.11)$$

where

$$y(t) = \text{actual output}$$
$$d(t) = \text{desired output}$$

This is known as the Wiener least mean-square-error criterion and the general filtering process is termed a Wiener filter. Considerable work on wavelet optimisation has been carried out for seismic or geophysical signal analysis for which the data show many similarities to those of surface-penetrating radar.

The Wiener filter can be adjusted for several situations and, in the case where the signal is modified by additive white Gaussian noise, the optimum filter is a matched filter which is a standard approach to conventional radar signal processing. A matched filter is mathematically identical to a correlation receiver and provides an output

$$H(t) = \int_{-\infty}^{\infty} e(\tau) \cdot s * (\tau + \tau_0 - t) d\tau \qquad (6.12)$$

where

$$
\begin{array}{ll}
e(\tau) & = \text{input signal} \\
s * (\tau - \tau_0 + t) & = \text{time reverse of the input signal} \\
\tau_0 & = \text{delay required to meet causality requirements}
\end{array}
$$

For this filter the output is maximum at $\tau_0 = t$.

An alternative realisation of the Wiener filter is the inverse filter. In this case the spectral response of the filter is the reciprocal of the signal spectrum, although in fact the extremes of the frequency range are limited to avoid excessive out-of-band energy degrading the output from the filter.

The frequency characteristic of an inverse filter is given by

$$H(f) = \frac{S * (f)}{N' + K|S(f)|^2} \qquad (6.13)$$

where

$$
\begin{array}{ll}
S * (f) & = \text{the Fourier transform of the time reverse of the signal} \\
N' & = \text{the noise power (halved)} \\
S(f) & = \text{signal spectrum} \\
K|S(f)|^2 & = \text{clutter power}
\end{array}
$$

The inverse filter results from the case where the clutter is largest and the matched filter results from the case where the noise predominates.

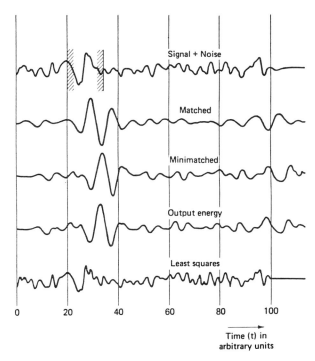

Figure 6.4   Methods of detection of a signal immersed in white noise (after Robinson and Treitel)

A comparison of the performance of several different configurations of filter is given in an excellent treatment by Robinson and Treitel (1984). The least-squares filter, optimised to provide a near delta-function output, was compared with a matched filter, a mini-matched filter and an output energy filter as shown in Figure 6.4.

It can be seen that the least-squares filter does not produce a high-amplitude output in the presence of noise, whereas the matched filter does. Note, however, that the output of the matched filter shows the effect of the different phase lag characteristic of the latter compared with the mini-matched and output energy filters. The least-squares filter does provide the optimum output in the case where the noise is significantly lower than the signal.

The selection of a suitable filter is therefore dependent upon the characteristics of the signal. As the sample data set is composed of signal, noise and clutter, the question of the stability of the filter must be considered as it is not possible to achieve infinite resolution of the filter output and unless certain restrictions are placed, both on the filter coefficients and on the characteristics of the sample data set, the filter will behave in an unstable manner.

As most of the processing of surface-penetrating-radar data is carried out using digital filtering, and mainly in the form of software algorithms, it is instructive to consider an example of those aspects of the design of a digital inverse filter which affect performance and stability.

If we assume that a sampled data set described by $d_0, d_1, d_2 \ldots d_n$ is the input to a filter f_t which then transforms the input to an output of unit magnitude at $t = 0$, then

$$h_t = \sum_{n=0}^{t} d_n f_{t-s} \tag{6.14}$$

for all positive values of t where

$$h_t = 1 \text{ for } n = 0$$
$$h_t = 0 \text{ for } n = 1 \text{ to } t$$

The required filter will be the time inverse of the input data series and will possess an infinite length of weighting coefficients or memory function.

Provided that

$$\sum |f_{t-n}|^2 \tag{6.15}$$

is finite the filter will be stable.

If the coefficients are derived using the z transform in order to represent the example in terms of unit delay, then the input to the filter is given by

$$D(z) = d_0 + d_1 z + d_2 z^2 \ldots \tag{6.16}$$

and the output

$$H(z) = 1$$

then

$$H(z) = D(z)F(z) \tag{6.17}$$

where

$$F(z) = \frac{H(z)}{D(z)} = \frac{1}{D(z)} \tag{6.18}$$

Hence

$$f_0 + f_1 z + f_2 z^2 \ldots = \frac{1}{d_0 + dz + d_2 z^2 \ldots} \tag{6.19}$$

If we consider an input with two terms d_0 and d_1 where $d_0 = 1$ and $d_1 = k$, the right-hand side is expanded by the Binomial theorem.

$$f_0 + f_1 z + f_2 z^2 \dots = 1 - kz + k^2 z^2 - k^3 z^3 \dots \tag{6.20}$$

Then equating coefficients

$$(f_0, f_2, f_3, f_4 \dots) = (1, -k, k^2, -k^3 \dots) \tag{6.21}$$

If k is less than 1, i.e. the input function is a minimum-delay wavelet, it can be seen that the values of the filter coefficient will converge to zero whereas, if the value of k is greater than 1, the series will diverge and the filter will be unstable. Thus the causal or one-sided filter is limited to processing minimum-delay wavelet functions.

This limitation can be overcome by generating a noncausal filter function. This can only be synthesised as, the signal having been recorded, there is no restriction on operating in real time. Hence the noncausal filter can operate in nonreal time with a set of coefficients that caters for minimum-delay and maximum-delay wavelet functions. A noncausal filter will have coefficients that extend equally either side of a nominal zero time. Alternatively, we can consider such a filter as possessing an inherent time delay. The time delay or lag of the filter can be optimised to provide the minimum error energy which, when normalised, equates to the average squared error e divided by the autocorrelation function of the desired output at the zero lag time.

When $e = 0$ the required and actual filter outputs are equal for all values of time and vice versa. The filter performance can be defined in terms of efficiency η as

$$\eta = 1 - e \tag{6.22}$$

For any filter the filter length and filter lag can be separately chosen to optimise η. However, if the filter length is larger than the separation between wavelets then the filter output will contain 'leaked' energy from any adjacent wavelet and will be suboptimum. The objective of optimum filtering is to generate an output shape (Gaussian, Delta function etc.) from an input wavelet which exists in a general noise background using the minimum-duration filter with a high value of η.

A relevant application of this approach is given by Ueno and Osumi (1984) who used the general method to produce a mixed delay filter to provide optimum matched filtering.

More recent developments in wavelet theory have introduced the concept of the wavelet transform (WT) which is relevant to the analysis of nonstationary signals such as the output from ultrawideband radar systems. Wavelet transforms are a class of transforms which decompose signals into a set of base functions or wavelets. These are obtained from a single elementary wavelet by expansion, contraction or shifting. The short-time Fourier transform (STFT) is the precursor of the wavelet transform (WT) and maps a signal onto a time–frequency plane as a time–frequency representation of the signal.

The Fourier transform is concerned with transformations applied to stationary signals, i.e. those signals whose properties do not evolve with time, for example sinewaves.

The Fourier transform of $x(t)$ is given by

$$F(\omega) = \frac{1}{2\pi} \int_{-\infty}^{\infty} x(t) \, e^{-j2\pi ft} dt \tag{6.23}$$

and the global domain is that of frequency. The Fourier transform is not well suited to sudden changes in time in a nonstationary signal, as the transform $F\{x(t)\}$ becomes spread out over the frequency domain. The STFT considers a signal over a limited time window $g(t)$ centred at time τ. Hence the STFT

$$F(\tau, \omega) = \frac{1}{2\pi} \int x(t) \cdot g * (t - \tau) \, e^{j2\pi ft} dt \tag{6.24}$$

Essentially the STFT introduces a frequency dependence with time by filtering the signal 'at all times' with a bandpass filter centred on each individual frequency and whose impulse response is that of the window function. The descriptive principle is similar to that of a score of a piece of music in which frequencies are 'played' in time.

The main limitation of the STFT is its inability to resolve more closely than the equivalent width of the bandpass filter, i.e.

$$\text{time-bandwidth product} = \Delta t \, \Delta f > \frac{1}{4\pi} \tag{6.25}$$

which is fundamentally related to the uncertainty principle.

The way of improving this limitation is to vary Δt and Δf in the time–frequency plane in order to obtain variable resolution. The STFT employs a constant value of $g(t)$ and exhibits a constant bandwidth. If Δf is varied so as to provide constant proportional bandwidth, i.e.

$$\frac{\Delta f}{f} = k \tag{6.26}$$

it is possible to achieve good time resolution at high frequencies while good frequency resolution at low frequencies. This is the basis of the continuous wavelet transform (WT) in which a prototype wavelet $h(t)$ serves as the basic wavelet and is scaled as appropriate; hence

$$h_a(t) = \frac{1}{\sqrt{|a|}} h\left(\frac{t}{a}\right) \tag{6.27}$$

and the transformed wavelet is given by

$$F_\omega(\tau, a) = \frac{1}{\sqrt{|a|}} \int_{-\infty}^{\infty} x(t) \cdot h * \left(\frac{t-\tau}{a}\right) dt \qquad (6.28)$$

It is usually more convenient to display the power-spectral density of a function rather than the real and imaginary components of its two-sided spectrum. The power-spectral density function of a normal Fourier spectrum has equivalence in what is termed a spectogram for the STFT or a scalogram for the WT.

The spectogram of a short-time Fourier transform is defined as

$$SPEC_x(t,f) = |STFT(t,f)|^2 \qquad (6.29)$$

and the scalogram of a Wavelet transform is defined as

$$SCAL_x(t,f) = |W\bar{T}(t,f)|^2 \qquad (6.30)$$

and both are convenient ways of viewing the signal in an analogous manner to the power spectral density.

Care should be exercised in the use of particular wavelet transforms, as there is an implicit assumption of linearity.

The spectrogram of the signals $C_1 x_1(t) + C_2 x_2(t)$ is not given by $SPEC_{x_1}(t,f) + SPEC_{x_2}(t,f)$ because the operation of squaring implies a quadratic superposition, i.e.

$$SPEC_x(t,f) = |C_1|^2 SPEC_{x_1} + |C_2|^2 SPEC_{x_2} + C_1 C_2 SPEC_{x_1 x_2} + C_2 C_1 SPEC_{2c_2} x \qquad (6.31)$$

The latter terms are known as interference terms and increase quadratically with the number of pairs of terms.

A range of wavelet transforms is available for analytic purposes and further details can be found in Hlawatch and Boudreaux-Bartels (1992).

The key feature of the various time–frequency representations such as the STFT, spectrogram, WT, scalogram, Gabor transform and Wigner distribution, is that they allow successful analysis of nonstationary signals in the form of either wavelets or chirp (linear FM) signals and permit much clearer target identification than conventional spectral-analysis techniques. An example of a Wigner distribution (Rioul and Vetterli, 1991) is given in Figure 6.5.

There is the distinct possibility of multiple echoes or reverberations in surface-penetrating radar. These can occur as a result of reflections between the antenna and the ground surface or within cables connecting the antennas to either the receiver or transmitter. The effect of these echoes can be considerably increased as a result of the application of time-varying gain. This can easily be appreciated

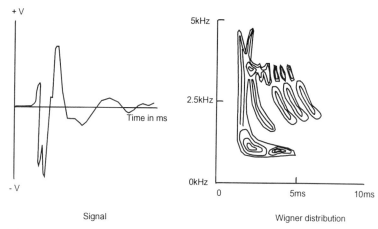

Figure 6.5   *Wigner distribution*

by considering the case of an antenna spaced at 0.5 m from the ground surface. Multiple reflections will occur every 1 m (twice the separation) and will reduce at a rate of equal to the product of the ground-surface reflection coefficient and the antenna reflection coefficient. This product is likely to be −15 dB; hence a series of echoes will occur every 3.3 ns reducing by 15 dB each time.

This problem may be partially overcome by suitable signal-processing algorithms which can be applied to the sampled time-series output from either time-domain or frequency-domain radars. The general expectation is that all the individual reflections will be minimum delay. This explanation is generally reliable because most reflection coefficients are less than unity; hence the more the impulse is reflected and rereflected the more it is attenuated and delayed. As a result the energy is concentrated at the beginning of the train of wavelets.

The simplest method of removing multiple reflections is by means of a filter of the form

$$F(z) = \frac{1}{1 + cz^n} \tag{6.32}$$

In essence this filter subtracts a delayed (by n) and attenuated (by c) value of the primary wavelet from the multiple wavelet train at a time corresponding to the arrival of the first reflection.

An alternative method relies on the use of filtering techniques applied to signals that have been combined by multiplication and by convolution. Such methods are termed homomorphic deconvolution filtering and rely on the fact that multiple reflections cause periodicity in the spectra of sampled data, as shown in Figure 6.6 (Randall and Lee, 1981).

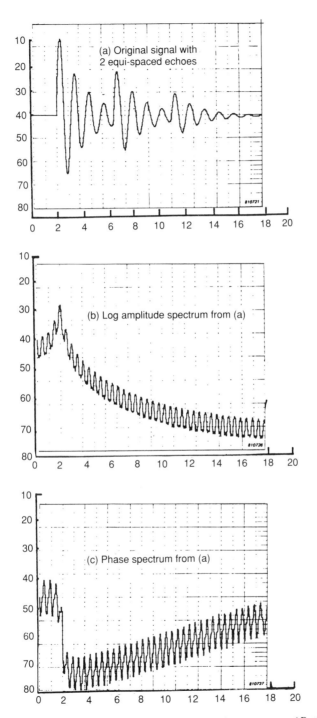

Figure 6.6   Multiple echoes and their amplitude and phase spectrum (Brüel and Kjaer)

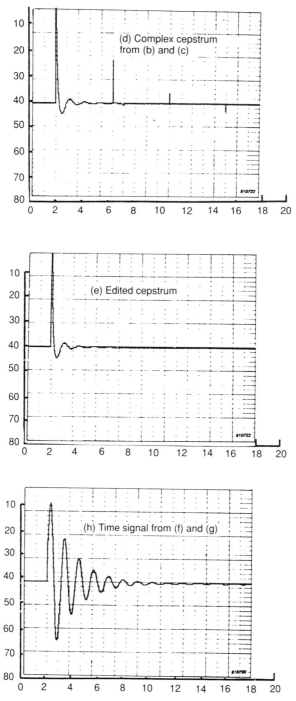

Figure 6.7   Cepstrum of Figure 6.6 and echo-removal process (Brüel and Kjaer)

The frequency spacing can be determined by taking the logarithm of the spectrum and then carrying out a spectral analysis of the new frequency series. This domain is known as the Cepstrum and is given by

$$C_c(\tau) = F'[c\log\{F(F(t))\}]\tag{6.33}$$

The essence of operation in the Cepstrum is that convolutions in the time domain are transposed to additions in the Cepstrum. As phase information is retained, it is possible to eliminate multiple echoes, i.e. wavelets, in the time domain by subtracting these in the Cepstrum and then inverse transforming to recover the original time series. The most significant difficulty of carrying out a complex logarithm is that the phase of the complex log must be a continuous function; hence discontinuities at intervals of 2π must be removed. An example of a Cepstrum filtering operation is shown in Figure 6.7 and it can be seen to be highly effective at multiple echo removal.

An alternative method of resolving overlapping echoes is based on the use of the multiple signal characterisation or MUSIC algorithm. The latter is a high-resolution spectral-estimation method and is used to estimate the received signal's covariance and then perform a spectral decomposition. Although computationally intensive, evaluation of the technique by Schmidt (1986) gave promising results.

(g) Target resonances

The application of analytic methods of target discrimination started with the application of Prony's method (1795) to target recognition. The basis of the technique is that every object will possess a unique resonant characteristic. Hence every target can be identified in terms of its resonant characteristic.

Van Blaricum and Mittra (1978) modelled a waveform by a series of exponentials in which the amplitude and delay constant are variable, i.e.

$$P(t) = \sum_i a_i \exp(\alpha_1 t)\tag{6.34}$$

A discretised version of the above gives

$$\alpha_n = \sum_{n=1}^{N/2} A_n \exp j\phi_n \exp(\alpha_n + j\omega_n)n\Delta t\tag{6.35}$$

where each parameter is given as follows:

$$A_n = \text{amplitude}$$
$$\phi_n = \text{phase}$$
$$\alpha = \text{damping factor}$$
$$\omega = \text{frequency}$$

In its basic form, Prony's method is inherently an ill-conditioned algorithm and is highly sensitive to noise, hence estimates of the number of poles present in the data may be prone to error.

It can, therefore, be understood that, for targets buried in a lossy medium, the high-frequency signal information is low and hence the signal-to-noise ratio is such as to make Prony processing very vulnerable. Indeed, Dudley (1988) points out that 'since all real data are truncated only approximations to the resonances are ever available even in the limit of vanishing noise'.

Recent developments have improved the robustness of the method as a means of detecting shallowly buried anti-tank or anti-personnel mines.

Work by Chan *et al.* (1981*b*) demonstrated the feasibility of the technique as a means of mine detection.

In general the received signal is represented as

$$C_r(t) = \sum_{n=1}^{N} a_i \exp(s_1 t) \tag{6.36}$$

where s_1 are the complex resonant frequencies in the complex frequency plane (*S*-plane), i.e. $s = \sigma + j\omega$.

A typical *s*-plane representation from various buried targets is shown in Figure 6.8.

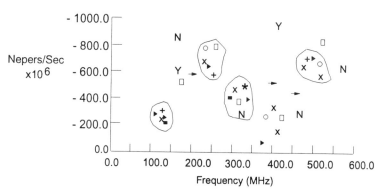

Figure 6.8   *S-Plane analysis of data*
Metallic anti-personnel mine-like Target 1 inch deep, 1 inch (25.4 mm) antenna [After Young]

There are two realisations of the Prony method, the classical and the eigenvalue method, and details of these are discussed by Chan *et al.* (1981*a*). The eigenvalue method was found to give better results, and by reference to eqn. 6.36 it is found that analysis of the wavelet resulting from the resonance is highly sensitive to the choice of the parameters. In addition, a wide dynamic range is needed to cater for both the early and late time portions of the wavelet.

Many of the signal-processing techniques applied to the data generated by surface-penetrating radar come from other disciplines such as geophysics and acoustics. There have been significant developments in ultra-wideband free-space radar technology and one goal of much research currently being carried out is the unique identification of a target from its ultra-wideband impulse response.

An appreciation of these techniques is valuable when considering their application to surface-penetrating-radar data, and the following Section reviews the main features of the approach.

Free-space scattering from resonant bodies has received considerable attention and the singularity-expansion method (SEM) is established as a means of target recognition. Much work in this area has been carried out by Webb (1984), Baum *et al.* (1991), Chen *et al.* (1986), Kennaugh (1981), Kennaugh and Moffatt (1965), Fok and Moffatt (1987) and Rothwell *et al.* (1987). The SEM suggest that the late time-scattered field of a target can be represented as the sum of excitation-independent natural resonance modes which depend on the detailed size and shape of the target.

An extension to the concept of nonspecific impulse illumination is that of discriminant functions designed to annihilate certain selected natural frequency constituents of the target response. The K pulse originated by Kennaugh (1981) is defined as that wavelet of minimum length which, when convolved with the target response, minimises or 'kills' all the natural modes in the resulting target response. Further developments in this area have resulted in the E-pulse (E = extinction) and the S-pulse (S = single mode).

The E-pulse is synthesised to minimise, when convolved with a band-limited late time target pulse response, all natural modes existing in that response.

As the scattered free-space far-field response of a conducting target can be expressed as

$$e(t) = \sum_{n=1}^{N} a_n \exp(\sigma_n t) \cos(\omega_n t + \phi_n) t > \tau_L \qquad (6.37)$$

where τ_L is the start point of the late time response, $e(t)$ is convolved with an extinction pulse wavelet $E(t)$ such that

$$c(t) = e(t) * E(t) = 0 \qquad (6.38)$$

The reader is referred to Baum *et al.* (1991) for details of the procedure for determining $e(t)$.

(h) Spectral-analysis methods

For those surface-penetrating-radar systems using FMCW, stepped-frequency or synthesised sources the receiver produces a frequency-domain signal which is then transformed to the time domain. Usually this is carried out using a fast Fourier transform (FFT). There are, however, several limitations of the FFT algorithm. While these are well documented, it is useful to review the capabilities of the FFT and alternative spectral-estimation methods. The reader is referred to a classic paper by Kay and Marple (1981) who detailed the following methods:

(i)   Conventional methods, Blackman Tukey, periodogram
(ii)   Modelling and parameter-identification approaches
(iii)   Rational transfer-function-modelling methods
(iv)   Autoregressive (AR) power-spectral-density estimation
(v)   Moving-average (MA) power-spectral-density estimation
(vi)   Autoregressive moving average (ARMA) power-spectral-density
        estimation
(vii)   Pisarenko harmonic decomposition
(viii) Prony energy spectral-density estimation
(ix)   Prony spectral-line estimation
(x)   Maximum-likelihood method (MLM)
(xi)   Maximum-entropy methods (MEM)

more recent developments include:

(xii)   Multiple signal classification (MUSIC)

The objective of this Section is not to describe all these methods in detail as there is adequate cover in the literature. However, highlighting of the underlying principles and capabilities of various spectral estimation methods should provide the reader with an introduction to the options which can be considered.

The FFT approach, while computationally efficient, (via the Cooley–Tukey or Blackman–Tukey algorithm), suffers from several limitations. First, it has a frequency resolution in Hz which is approximately equal to the reciprocal of the sample window duration in seconds. Secondly, the action of sampling the data for a defined time or 'window' causes leakage of energy from the main lobe of a spectral response into the sidelobes. As most data from surface penetrating radar occur in short data sequences, the windowing effect can be particularly difficult to counter while at the same time maintaining resolution. It should be noted that when the signal-to-noise ratio is low the results of the conventional FFT approach are comparable with those of the more modern spectral-estimation techniques.

In general the modelling approach to spectrum analysis is carried out in three stages: first, the selection of an appropriate time series model; secondly, an estimation of the parameters of the assumed model using either the sampled data or the autocorrelation lags; and finally, substitution of the estimated model parameters into the theoretical power spectral density function. The selection of the time-series model is governed by estimation and identification methods of linear systems theory. The modelling approach enables assumptions to be made concerning the time series outside the measurement window and hence eliminates the problems associated with windowing functions.

Three approaches to non-Fourier spectral estimation methods are: autoregressive (AR), moving average (MA) and autoregressive moving average (ARMA).

The autoregressive or feedback model is described by

$$y_n = a_0 x_n - a_1 y_{n-1} - a_2 y_{n-2} \ldots a_N y_{n-N} \tag{6.39}$$

where n is the sample variable, x is the input and y is the output variable. In terms of the z transform the

$$y_n = \frac{1}{a_0 + a_1 z + a_2 z^2 + a_3 z^3 \ldots a_N z^n} x_n \tag{6.40}$$

As in general the denominator can be considered to be a polynomial denoted by $A_n(z)$, then

$$Y_{(z)} = \frac{1}{A_n(z)} X(z) \tag{6.41}$$

For the AR model it is necessary to determine algorithms to solve the power-spectral representation of eqn. 6.41. As $z = e^{j\omega}$ it can be appreciated that as

$$F(\omega) = |Y(\omega)| \tag{6.42}$$

then

$$F(\omega) = \frac{1}{|A(\omega)|^2} |X(\omega)|^2$$

There are two methods of solving the AR method, the Yule–Walker and Burg methods. The Yule–Walker method tends to emphasise the spectral aspects at the expense of peak definition. The method originated by Burg is based on a technique known as maximum entropy. The entropy or unpredictability of a time series is proportional to the integral of the log of its power spectrum. An estimate of the power spectrum produced by the MEM method corresponds to the least predictable time series consistent with the autocorrelation function lags. The AR method tends to define the peak spectral response at the expense of spectrum shape.

The moving average (MA) or feedforward model is described by

$$y_n = b_0 x_n + b_1 x_{n-1} + b_2 x_{n-2} + \ldots b_n x_{n-N} \tag{6.43}$$

where y_n represents the output of the linear convolutional filter $(b_0, b_1, b_2, \ldots, b_n)$ for the input x_n. The output of the MA model is a linear combination of present and past values of the system input sequence x_n. When the input is white noise, the output is called a moving average process of order n, i.e. MA(n).

In terms of the z transform

$$y_n = \left(b_0 + b_1 z + b_2 z^2 + b_3 z^3 \ldots b_N z^N\right) x_n \tag{6.44}$$

The polynomial on the right-hand side of eqn. 6.44 can be expressed as $B_n(z)$, hence

$$Y(z) = B_n(z) x(z)$$

In a similar manner to the AR model, the spectral representation is given by

$$F(\omega) = |B_n(\omega)|^2 |X(\omega)|^2$$

The MA approach yields a reasonable approximation to the spectrum but tends to offer similar performance to the Blackman–Tukey PSD.

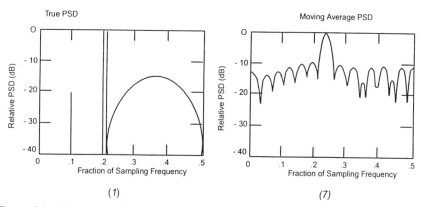

Figure 6.9   *Illustration of various spectra for the same 64-point sample sequence (Kay and Marple)*
1. True PSD
7. Moving-average PSD

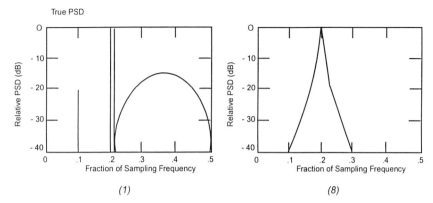

Figure 6.10 *Illustration of spectra for the same 64-point sample sequence (Kay and Marple)*

The ARMA model evidently contains both AR and MA elements and it can be shown that

$$F(\omega) = \frac{|B_n(\omega)|^2}{|A_N(\omega)|^2} |X(\omega)|^2$$

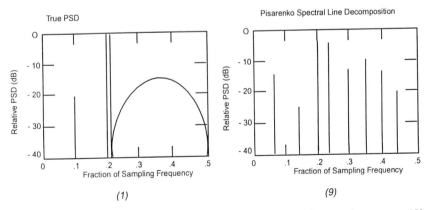

Figure 6.11 *Illustration of various spectra for the same 64-point sample sequence (Kay and Marple)*
1. True PSD
9. Pisarenko spectral line discrimination

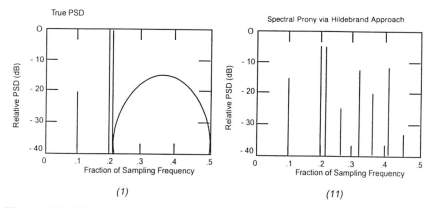

(1)   (11)

Figure 6.12   Illustration of various spectra for the same 64-point sample sequence (Kay and Marple)
    1 True PSD
    11 Spectral Prony via Hildebrand approach

Kay and Marple compared results of spectral-estimation methods for various approaches and these are shown in Figures 6.9–6.13.

An alternative approach is that given by the MUSIC algorithm which produces a spectral estimation from the autocovariance matrix via eigenvector-value decomposition. An example of the resolution of a pair of sinusoidal waveforms is shown in Figure 6.14 which compares various results from different methods.

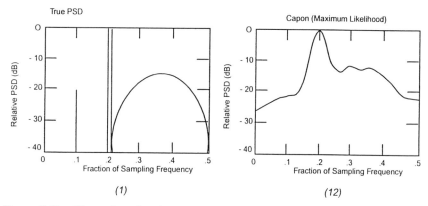

(1)   (12)

Figure 6.13   Illustration of various spectra for the same 64-point sample sequence (Kay and Marple)
    1 True PSD
    12 Capon maximum likelihood

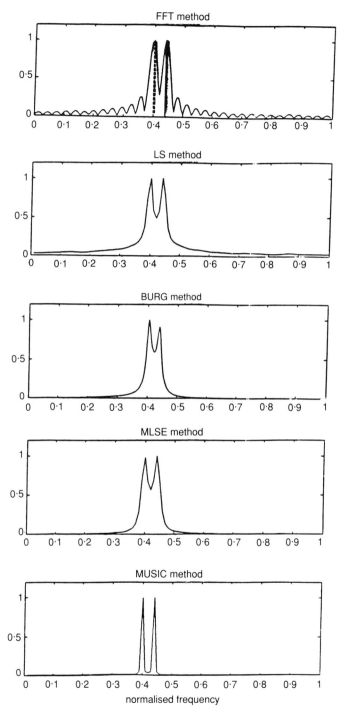

Figure 6.14 Comparison of MUSIC method of spectral analysis with a range of alternatives

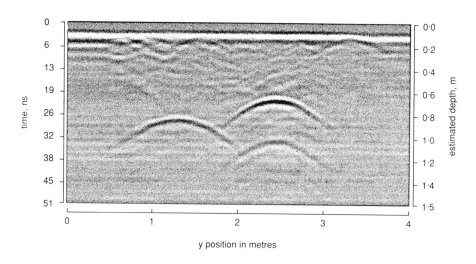

Figure 6.15   Gain profile of 5 dB/m (correct) (courtesy ERA Technology)

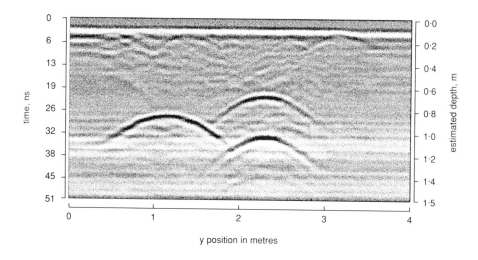

Figure 6.16   Gain profile of 15 dB/m (excessive) (courtesy ERA Technology)

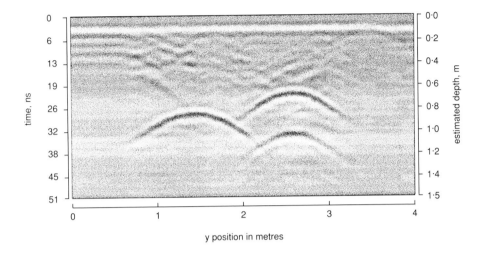

Figure 6.17   Lowpass-filtered data (courtesy ERA Technology)

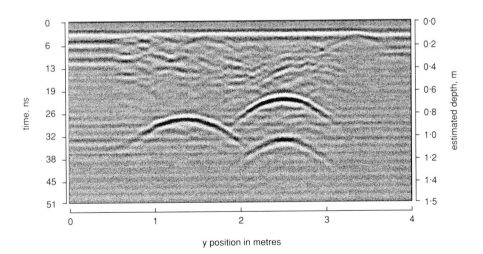

Figure 6.18   Highpass-filtered data (courtesy ERA Technology)

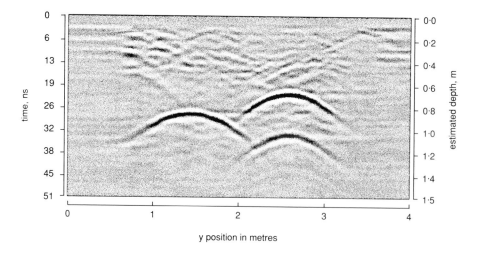

Figure 6.19   Average-removal data (courtesy ERA Technology)

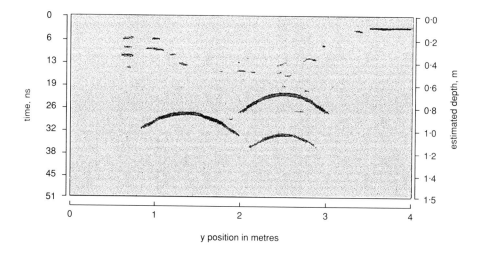

Figure 6.20   Absolute-thresholded data (courtesy ERA Technology)

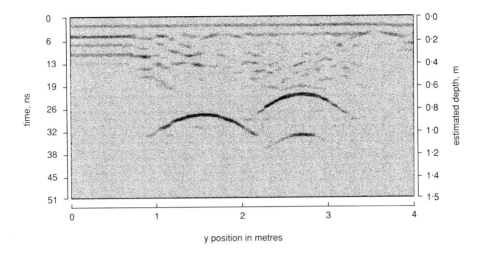

Figure 6.21   Mean-value thresholded data (courtesy ERA Technology)

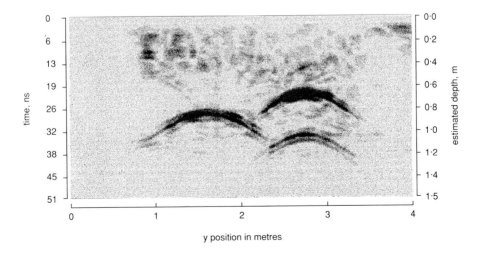

Figure 6.22   Peak envelope applied to data (courtesy ERA Technology)

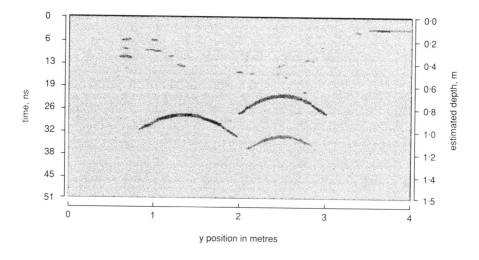

Figure 6.23   Peak-indication algorithm applied to data (courtesy ERA Technology)

(*h*) *Examples of processing techniques*

A set of examples of various methods of A-Scan processing is shown in Figures 6.15–6.23. They are shown as grey-scale B-Scans to illustrate the effect of applying various algorithms on a radar image of three plastics pipes buried in wet sand at depths up to 1.3 m.

6.3 B-scan processing

If we consider an ensemble set of time samples comprising a B-Scan, there are a number of approaches to signal processing which can be considered. The data, even after optimal A-Scan processing, are usually unfocused in that the spatial antenna response is convolved with the target spatial response as shown in Figure 6.24. For planar features such as the interfaces between layers in a road or the water table in a geophysical survey it may be considered uneconomic to focus the data. However, for small or complex objects it may be valuable to reduce the effect of the spatial smearing caused by the antenna. This Section considers several approaches to B-Scan processing, the chief of which is based on migration of the data.

(*a*) *Migration*

The migration process essentially constructs the target reflector surface from

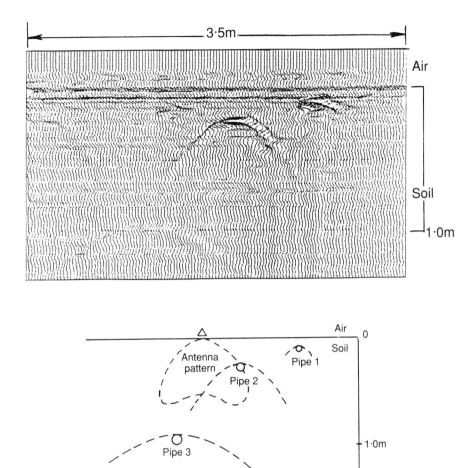

Figure 6.24   Convolution of antenna pattern with target (courtesy ERA Technology)

the record surface. The migration technique has been much developed in acoustic, seismic and geophysical engineering and was originally developed in two-dimensional form by Hagedoorn (1954). More recent developments employ wave-equation methods such as Kirchoff migration, finite-difference migration and frequency wave-number migration. An excellent tutorial paper is given by Berkhout (1981) and there are also papers by Schneider (1978) and by Robinson and Treitel (1980). In essence the problem of data focusing can be considered from the point of view of a source of radiation, i.e. a point reflector and the measured wavefront, as shown in the simulated data of Figure 6.25.

A relatively straightforward geometric approach can be used in the two-

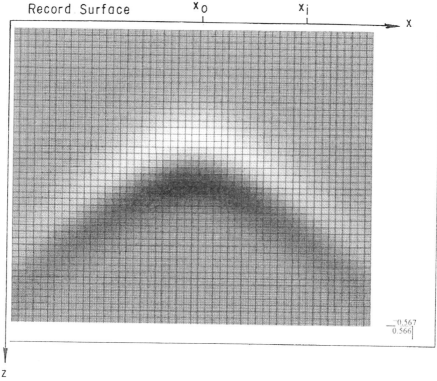

Figure 6.25   Migration of a wavefront from a point source (calculated)

dimensional case of a material with known constant velocity. If the measured time to the point reflector is t then the distance to the point reflector is given by $z = vt/2$. At any position along the x axis the distance z is also given by

$$z_i = \sqrt{\left((x_i - x_0)^2 + z_0^2\right)} \qquad (6.45)$$

This equation shows that the measured wavefront appears as a hyperbolic image or a curve of maximum convexity. The geometric migration technique simply moves or migrates a segment of an A-scan time sample to the apex of a curve of maximum convexity. The hyperbolic curve needs to be well separated from other features and a good signal-to-noise ratio is needed.

An alternative method is known as the maximum-convexity migration and is computationally intensive but straightforward in concept. The method assumes a semi-hyperbolic maximum-convexity function and sums the value of each separate A-scan at the point at which it intersects the semi-hyperbolic focus over the ensemble data set. All in phase energy adds in phase whereas noncoherent energy is usually out of phase and tends to zero. The method assumes a constant velocity and must, of course, be adjusted as a function of depth. It also requires

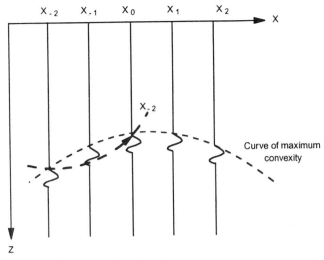

Figure 6.26   Maximum-convexity migration

that each and every possible point be examined. Care must also be taken over the window after which the summation is carried out. The process is shown diagrammatically in Figure 6.26.

The alternative process to maximum-convexity migration is wavefront migration. In this process every section of the individual A-scans comprising a B-scan is mapped to a migrated B-scan. The mapping function is, in the constant-velocity case, a semicircle of radius equal to the depth of the section.

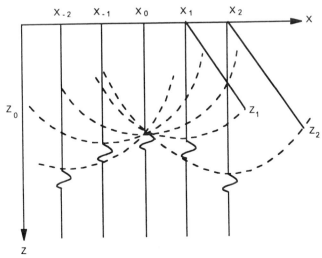

Figure 6.27   Wavefront migration

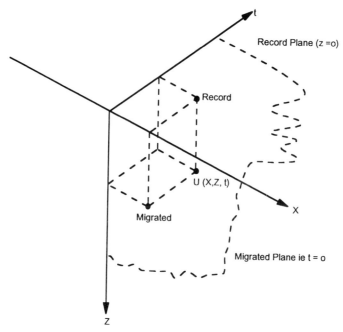

Figure 6.28   Wave fields at reflector

This process is repeated for each section of the individual A-scan and for all A-scans, and the resultant output is the superposition of all the migrated sections. The process is shown in Figure 6.27 and it can be seen that wavelets from neighbouring A-scans add constructively to produce a focused reflection point.

The methods described in the previous paragraphs illustrate the general principle of migration but are relatively unsophisticated methods which are limited to the constant-velocity situation and assume a point-source reflector.

Alternative and more comprehensive techniques use wave equation methods. Berkhout (1981) showed that wave-field extrapolation techniques are based on three methods: the Kirchoff summation approach, the plane-wave method ($k - f$ method) and the finite-difference method.

Although the methods described by Berkhout are founded in acoustic propagation theory which is based on a scalar wave equation, they are still relevant to surface-penetrating-radar data even though the latter are derived from electromagnetic soundings. The general process of imaging of such data consists of two operations. The first consists of a wave field extrapolation whereby, using a scalar wave equation, the recorded data are transformed into a new data series which represent simulated recordings at new positions of the measurement plane. The second operation of imaging generates the data around the zero time-travel position of the simulated recording related to planes within the B-scan or C-scan.

The wave field at the ground or material surface is given by $\nu(x, 0, t)$ and the A-scan observed at the surface is given by $\nu(x, z, t)$. Note that time is considered as positive and the value $\nu(x, 0, t)$ represents all time information along the line x.

The wave field at the reflector is given by $\nu(x, z, 0)$ as shown in Figure 6.28.

The procedure for evaluating the reflector surface is given by Robinson and Treitel (1980) using the Fourier transform approach as frequency–wave-number domain migration. The method is carried out in three stages, the first being given by

$$F(k_x, \omega) = \int_{-\infty}^{\infty} \int_{-\infty}^{\infty} \nu(x, 0, t) \exp\{-t(\omega t + k_x x)\} dx dt \qquad (6.46)$$

where

$$k_x = \text{horizontal wavenumber}$$
$$k_z = \text{vertical wavenumber}$$
$$\omega = \text{angular frequency}$$
$$c = \text{propagation velocity of the medium}$$

From this a two-dimensional Fourier transform with respect to k_x and k_z is determined by

$$A(k_x, k_z) = F\left[k_x, ck_z\sqrt{\left\{1 + \left(\frac{k_x}{k_z}\right)^2\right\}}\right] \frac{c}{\sqrt{\left\{1 + (k_x/k_z)^2\right\}}} \qquad (6.47)$$

The reflector can then be derived as the inverse transform of $A(k_x, k_z)$ where

$$\nu(x, z, 0) = \frac{1}{4\pi^2} \int_{-\infty}^{\infty} \int_{-\infty}^{\infty} A(k_x, k_z) \exp\{j(k_x x + k_y z)\} dk_x dk_z \qquad (6.48)$$

The frequency-wave-number approach described above was first developed by Stolt and then used by Hogen (1988) to migrate ground-penetration-radar data. However, Fisher and McMechan (1992) comment that the method is difficult to implement efficiently when the propagation velocity varies. Fisher compared the Kirchoff diffraction method with the reverse-time migration approach which has the advantage of taking into account arbitrary velocity variations.

This is particularly important as, when migration velocities are chosen which are lower than the correct velocity, diffraction 'tails' will extend downwards from the reflector location. When the migration velocity is too high the tails will extend upwards from the reflector location.

The scalar wave equation is given by Mitchell (1969):

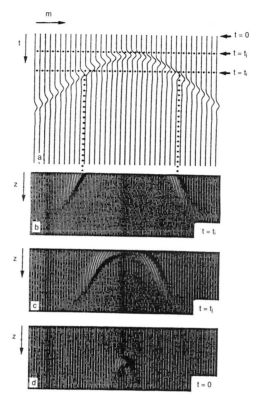

Figure 6.29   Reverse-time migration (courtesy Fisher)

$$\frac{\partial^2 U}{\partial x^2} + \frac{\partial^2 U}{\partial z^2} = \frac{1}{\nu^2(x,z)}\frac{\partial^2 U}{\partial t^2} \qquad (6.49)$$

where

U is the scalar wave field

$\nu(x,z)$ is the local propagation velocity.

The wave equation can be solved via an explicit second-order central finite-difference method.

These methods have been developed by Claerbout (1985) and McMechan (1983) and an example of reverse time migration is shown in Figure 6.29.

Computer programs to carry out migration are reported in the literature and Siggins (1990) includes an example of a convolution method written in C language.

(b) Synthetic-aperture processing

As the antenna of a surface-penetrating radar moves, the process can be considered analogous to the antenna of a synthetic-aperture radar (SAR). However, the attenuation of the earth material limits the improvement that can be obtained by, effectively, windowing the synthetic aperture.

Synthetic-aperture processing requires measurements to be made at a number of antenna positions (i.e. a B-scan) and then combined to simulate a narrow beam. The improvement that can be gained is related to the length of the synthetic aperture.

In general the resolution in either x or y direction of an antenna used in a SAR is given by

$$\delta_x = 2Z\theta_x \text{ or } \delta_y = 2Z\theta_y$$

where

$$Z = \text{depth of the target}$$
$$\theta = \text{beamwidth}$$

As most surface-penetrating-radar antennas are dielectrically loaded dipoles, the beamwidth is in the order of $\pi/2$ radians hence the resolution is poor.

If the equivalent antenna phase front is considered to be a plane wave and is uncorrected, the SAR is termed unfocused and the effective aperture is given by Skolnik (1970) as $R\theta$; hence the resolution is given by

$$\delta_x = \sqrt{(z\lambda)}/2 \tag{6.50}$$

As most surface-penetrating-radar antennas are not used in plane-wave conditions and are most likely to be operated in the near-field and Fresnel region, it is necessary to provide an amplitude and time-delay correction for each range.

In the Fresnel region the resolution in free space would be given by

$$\delta_x = \frac{z}{2} \tag{6.51}$$

However, the effect of attenuation is to reduce the resolution and, assuming an inverse fourth power relationship of received power with depth, Daniels *et al* (1988) gives

$$\delta_x = 4z\sqrt{\frac{\ln 2}{2 + \alpha z}} \tag{6.52}$$

where α is the attenuation coefficient in decibels per metre.

It can be concluded therefore that horizontal resolution improves as the material attenuation increases, whereas the resolution of a synthetic aperture would degrade under conditions of increasing attenuation.

In most situations the target is close, at least in terms of average wavelength, to the measurement plane and thus some of the methods described in Section 6.2 become more relevant. Holographic synthetic-aperture imaging methods have been developed for the single-frequency case

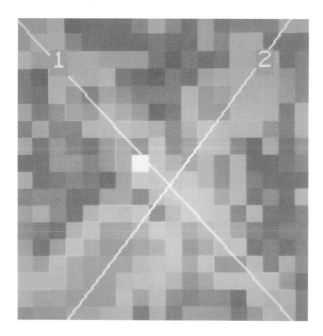

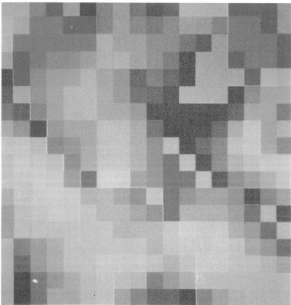

Figure 6.30   Holographic image focused to the object plane (courtesy of Junkin)
    a Amplitude
      1 = Plastic pipe at 0.3 m
      2 = Metal pipe at 0.3 m
    b Phase

by Junkin and Anderson (1986) (see Figure 6.30) and for the multifrequency case by Osumi and Ueno (1988).

The field strength at any point on the record surface can be derived from Fresnel–Kirchoff diffraction theory for a spherical wave in lossy dielectric and is given (Osumi and Ueno, 1988) by

$$v(x_0, t) = \frac{1}{2\pi v} \iint_s p(x) \frac{z(l_t + l_r)}{(l_t l_r)} \exp\{-\alpha(l_t + l_r)\} U'\left(t - \frac{l_t + l_r}{v}\right) dxdy \quad (6.53)$$

where

$v =$ propagation velocity

$\alpha =$ attenuation coefficient

$l_t =$ transmit path distance

$l_r =$ receive path distance

$U' =$ time derivative of the input pulse waveform $u(t)$

The image reconstruction implies the estimation of reflectance $p(x)$ from the transmitted signal and a set of recorded returns. The holographic process correlates the set of recorded returns and test functions and gives an image function

$$b(x) = \iint \int_{-\infty}^{\infty} U(x_0, t)h(x - x_0, t) dt dx_0 dy_0 \quad (6.54)$$

The key to the imaging process is the optimisation of the test function, and the various references deal with this process in greater detail.

SAR methods have been applied to single-frequency, impulse and FMCW radars and Yamaguchi *et al.* (1990) describe the process of determination of the object reflection distribution function. For an FMCW radar a B-scan data set is recorded.

The process of producing the target distribution comprises Fourier transforming each A-scan and multiplying this by the Fourier transform of the inverse propagation function. The product is then inverse transformed. A typical holographic SAR image (Skolnik, 1970) is shown in Figure 6.30.

6.4 Image processing

An alternative method of B-scan processing is based on image-processing techniques. Here it is assumed that the basic-level processing has been carried out and the B-scan is presented either as a grey scale or colour-coded image as shown in Figures 6.31 and 6.32. As radar data are bipolar in contrast to data derived from optical sensors, it is normal to encode the amplitude of the radar data in a defined manner. Hence the amplitude scaling for grey scale would assign most negative values to black and most positive to white. Colour coding is more complex and it is found preferable to limit the range of the colour palette

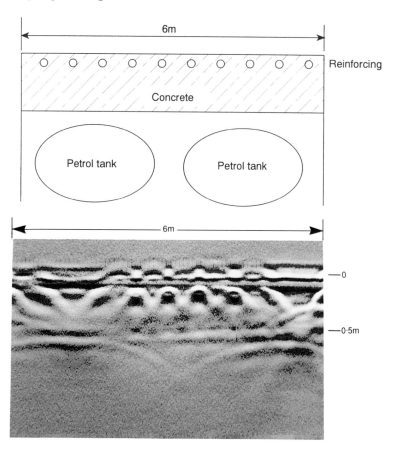

Figure 6.31   B-scan in grey scale (courtesy ERA Technology)

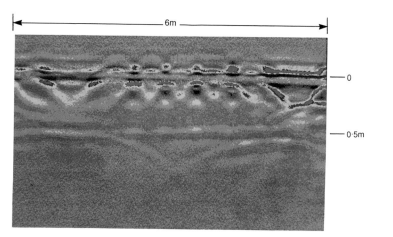

Figure 6.32   B-scan in colour (courtesy ERA Technology)

as the eye is more sensitive to intensity within a particular colour range rather than a wide range of colours which can be found confusing.

The general method of image processing is well established. Normally a two-dimensional mask filter, whose coefficients are set to those of the image of the target (for which analysis is required), is convolved with the data. Alternatively, a three-dimensional data set is recorded for subsequent processing and two-dimensional data images in any orthogonal plane are produced.

Image-processing techniques involve filtering operations which require fast execution of two-dimensional convolution algorithms. These can be divided into four categories:

(i)   lowpass
(ii)   highpass
(iii) edge detection
(iv)   template matching

All four types are commonly implemented by convolving the input data with a two-dimensional array of filter coefficients called the kernel. Each type performs a different function in the image-enhancement–restoration process and multiple types can be used to improve the image visibility. The most common convolution kernels are 3×3 and 5×5 and sets of commonly used kernels are listed below.

(*a*) 3×3 highpass
   Highpass filter (divisor = 1)

$$
\begin{array}{rrr}
-1 & -1 & -1 \\
-1 & 9 & -1 \\
-1 & -1 & -1
\end{array}
$$

(*b*) 3×3 lowpass
Lowpass filter (divisor = 9)

$$
\begin{array}{ccc}
1 & 1 & 1 \\
1 & 1 & 1 \\
1 & 1 & 1
\end{array}
$$

(*c*) 3×3 Laplacian
Laplacian filter (divisor = 1)

$$
\begin{array}{rrr}
-1 & -1 & -1 \\
-1 & 8 & -1 \\
-1 & -1 & -1
\end{array}
$$

(*d*) 3 × 4 vertical line enhancement
Vertical line enhancement (divisor = 1)

$$
\begin{array}{rrr}
-1 & 2 & -1 \\
-1 & 2 & -1 \\
-1 & 2 & -1
\end{array}
$$

(*e*) 3 × 4 horizontal line enhancement
Horizontal line enhancement (divisor = 1)

$$
\begin{array}{rrr}
-1 & -1 & -1 \\
2 & 2 & 2 \\
-1 & -1 & -1
\end{array}
$$

Lowpass filtering smooths an image so that large objects are transformed into homogeneous zones and small objects are reduced in intensity and/or merged into the larger regions. In the process, edges are reduced in intensity so that the details of an object are lost and objects in close proximity to each other are combined.

Highpass filtering performs the opposite function. Here, fine details that might be missed in the original image are increased in intensity. Note that the same mathematical operation as that for the lowpass filter is performed here; only the coefficients in the kernel have been changed.

Edge detection (or extraction) amplifies abrupt changes in intensity of an image and removes all other information. The three basic kinds of edge detection are the Laplacian, the Prewitt, and the Sobel operators. The Sobel and Prewitt operators have the advantage of providing edge-direction information. However, they require two or more passes, one for each edge direction. On the other hand, the Laplacian operator is isotropic, i.e. it extracts the image edges from all directions. Edge detectors provide the same type of information as highpass filters, (i.e. they amplify details in an image) but are more amenable to post filtering; small changes in intensity can be eliminated from further processing by comparing the output of the edge detector with a threshold value and discarding pixel values below the threshold.

Template matching, or matched filtering, convolves the image with a kernel that contains a pattern to be detected in the data. Template matching is also often used in texture and pattern recognition to determine which parts of an image are most likely to contain targets. In processing radar data, several important factors must be considered. Firstly, unlike image data, radar data are bipolar; hence an initial operation to convert radar data to unipolar data is necessary. This can be achieved by either converting radar data to absolute data or thresholding at zero to give positive going or negative going data only. Secondly, the radar image does not correspond to geometric patterns and it will be necessary to identify suitable kernel coefficients which are appropriate to the radar image pattern rather than a geometrical model.

Further development of radar image processing by means of the Hough transform is reported by Kaneko (1991).

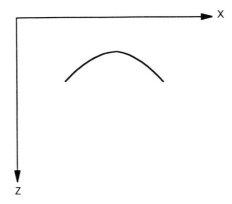

(a) Hyperbola in X, Z Space

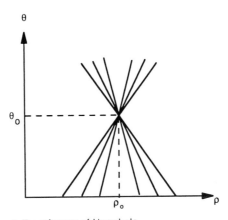

(b) Hough Transformer of Hyperbola

Figure 6.33   *Hough transform of hyperbola*
a Hyperbola in X, Z space
b Hough transformer of hyperbola

Essentially, the Hough transform transforms a line in image space to a point in polar-co-ordinate space. A straight line may be described by the equation

$$\rho = x \cos \theta + y \sin \theta$$

and the Hough transform becomes a point in (θ, ρ) space.

A hyperbolic reflection transforms to a set of lines as shown in Figure 6.33.

The method proposed by Kaneko (1987) extracts lines and curves, such as the hyperbolic reflection generated by a pipe, or pipes, from the radar image by means of a two-stage process. The first Hough transform uses the first derivative of edge contours to provide an initial fast estimate of the wanted target. The second part of the process uses a second derivative of the edge contours to improve the accuracy of the Hough transform.

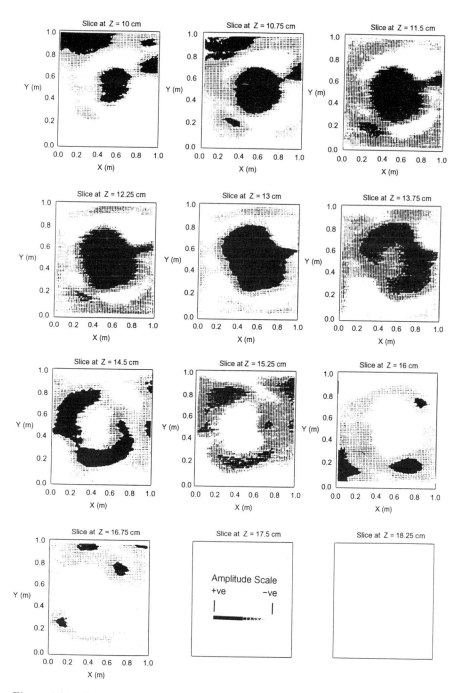

Figure 6.34   Set of C-scan depth slices of a buried minelike target (courtesy ERA Technology)

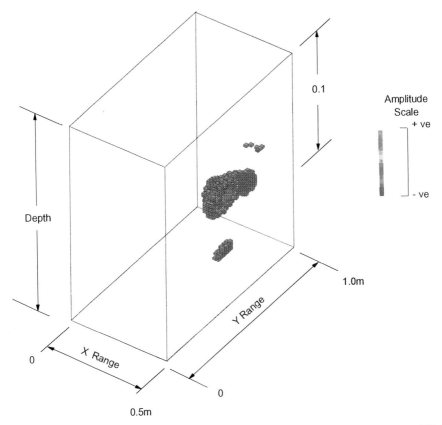

Figure 6.35   Unfocused radar image of a buried minelike target (courtesy ERA Technology)

6.5 C-scan processing

Although a C-scan is essentially an x, y plane at a selected value of z or range of values of z, many of the previous processes described in Section 6.4 can be applied. It is, however, important to focus the data, otherwise spurious features will appear at depths below the target. It is also important to ensure that adequate allowance is made for the sampling range, otherwise the familiar problems associated with windowing a data set will occur. Typical examples of C-scan processing are shown in Figure 6.34 and the radar image generated from a buried plastic mine-like target is shown in Figure 6.35.

6.6 Summary

The selection of suitable signal-processing methods must start from a clear appreciation of the modulation technique and the likely form of the received wavelet. The main initial objective is to select suitable processing to optimise the wavelet output in terms of each individual A-scan sample time series. If the subsequent objective is to generate an image, it is reasonable to consider some type of 'spiking' filter. If, however, the objective is to classify the wavelet, i.e. by Prony processing, then 'spiking' filters are not appropriate. Consideration can also be given to the removal of multiple reflections. Once the A-scan data are optimised, processing methods based on B-scan data sets can be considered. Again if the objective is image 'spiking' or focusing a number of migration or synthetic aperture methods are available, each of which is more or less tolerant to variations in propagation conditions. Where this method is not preferred, image pattern-recognition techniques based on standard image-processing methods or template maching can be used. Transforms such as Hough or neural-network techniques can also be considered.

6.7 References

BAUM, C. E., *et al.* (1991): Singularity expansion method. *Proc. IEEE*, **79**, (10), 1481-1492

BERKHOUT, A. J. (1981): Wave field extrapolation techniques in seismic migration, a tutorial. *Geophysics*, **46**, 1638–1656

CALDECOTT, R., YOUNG, J. D., HALL, J. P., and TERZUOLI, A. J. (1985): An underground obstacle detection and mapping system. Electro Science Laboratory, Ohio State University, report EL–3984

CHAN, L. C., PETERS, L., and MOFFATT, D. L. (1981*a*): Improved performance of a subsurface radar target identification system through antenna design. *IEEE Trans.*, **AP–29**, 307–311

CHAN, L. C., MOFFATT, D. L., and PETERS, L. (1981*b*): Subsurface radar target imaging estimates. *IEEE Trans.*, **AP–29**, 413–417

CHEN, K. M., NYQUIST, D. P., ROTHWELL, E. J., WEBB, L. L., and DRACHMAN, B. (1986): Radar target discrimination by convolution of radar returns with extinction pulses and single-mode extraction signals. *IEEE Trans.*, **AP–34**, 896–904

CLAERBOUT, J. F. (1985): *Imaging the earth's interior*. Blackwell Scientific Publications, Palo Alto

DANIELS, D. J., SCOTT, H. F., and GUNTON, D. J. (1988): Introduction to subsurface radar. *IEE Proc. F*, **135**, (4), 278–321

DUDLEY, D. G. (1988): Progress in identification of electromagnetic systems. *IEEE Antennas Propag. Newslett.*, August, pp. 5–11

FISHER, E., and MCMECHAN, G. A. (1992): Examples for reverse-time migration of single-channel, ground-penetrating radar profiles. *Geophysics*, **57**, (4), 577–586

FOK, F. Y. S., and MOFFATT, D. L. (1987): The K-pulse and E–pulse. *IEEE Trans.*, **AP–35**, (11), 1325–1326

HAGEDOORN, J. G. (1954): A process of seismic reflection interpretation. *Geophys. Prospect.*, **2**, 85–127

HLAWATSCH, F., and BOUDREAUX-BARTELS, G. F. (1992): Linear and quadratic time-frequency signal representations. *IEEE Signal Process. Mag.*, **9**, 21–69

HOGAN, G. (1988): Migration of ground penetrating radar data. Abstracts of 59th annual international meeting on *Social Exploitation of Geophysics*, 345–347

JUNKIN, G., and ANDERSON, A. P. (1986): A new system for microwave holographic imaging of buried services. Proceedings of 16th *European Microwave Conference*, 720–725

KANEKO, T. (1991): Radar image processing for locating underground linear objects. *IECIE Trans.*, **74**, (10), 3451–3459

KAY, S. M., and MARPLE, S. L. (1981): Spectrum analysis – a modern perspective. *Proc. IEEE*, **69**, 1380–1419

KENNAUGH, E. M. (1981): The K-pulse concept. *IEEE Trans.*, **AP–29**, (2), 327–331

KENNAUGH, E. M., and MOFFATT, D. L. (1965): Transient and impulse response approximations. *Proc. IEEE*, **53**, (8), 893–901

MCMECHAN, G. A. (1983): Migration by extrapolation of time dependent boundary values. *Geophys. Prospect.*, **31**, 413–420

MITCHELL, A. R. (1969): *Computational methods in partial differential equation.* John Wiley

OSUMI, N., and UENO, K. (1988): Detection of buried plant. *IEE Proc. F*, **135**, (4), 330–342

PRONY, G. R. B. (1795): Essai experimental et analytique, etc. *J. de l'Ecole Polytechnique Paris*, **1**, (2), 24–76

RANDALL, R. B., and LEE, J. (1981): Cepstrum analysis. *Bruel & Kjaer Tech, Rev.*, (3)

RIOUL, O., and VETTERLI, M. (1991): Wavelets and signal processing. *IEEE Signal Process. Mag.*, **8**, 14–38

ROBINSON, E. A., and TREITEL, S. (1980): *Geophysical signal analysis.* Prentice–Hall

ROTHWELL, E. J., CHEM, K. M., and NYQUIST, D. P. (1987): Extraction of the natural frequencies of a measured response using E–pulse techniques. *IEEE Trans.*, **AP–35**, 715–720

SCHMIDT, R. O. (1986): Multiple emitter location and signal parameter estimation. *IEEE Trans.*, **AP–34**, (3), 276–280

SCHNEIDER, W. A. (1978): Integral formulation for migration in two and three dimensions. *Geophysics*, **43**, 49–76

SIGGINS, A. F. (1990): Ground penetrating radar in geotechnical applications. *Explor. Geophys.*, **21**, (3–4), 175–186

SKOLNIK, M. I. (1970): *Introduction to radar systems, 2nd edn.* McGraw-Hill

UENO, K., and OSUMI, N. (1984): Underground pipe detection based on microwave polarisation effect. Proceedings of international symposium on *Noise and Clutter Rejection*, Tokyo, Japan, 673–678

VAN BLARICUM, M., and MITTRA, R. (1978): Problems and solutions associated with Prony's method for processing transient data. *IEEE Trans.*, **AP–26**, (1), pp. 174–182

WEBB, L. (1984): Radar target discrimination using K–pulse from a 'fast' Prony's method. PhD dissertation, Department of Electrical Engineering & Systems Science, Michigan State University

YAMAGUCHI, Y., SENGOKU, N., and ABE, T. (1990): FM–CW radar applied to the detection of buried objects in snowpack. Proceedings of international symposium on *Antennas and Propagation - Merging Technologies for 90s*, Dallas, TX, USA, 738–741

Chapter 7

Applications

7.1 Introduction

The objective of this chapter is to illustrate, by means of special invitations, the range of applications of surface penetrating radar. Accordingly, a number of organisations and individuals were invited to submit a short contribution on their particular area of technical interest. While the style of contributions highlights the individuality of each, the variation serves to emphasise the breadth and diversity of current applications of surface-penetrating radar.

Each individual contribution is separately acknowledged and the author is most grateful for the support of all the contributors.

There are an increasing number of applications and organisations working on surface-penetrating radar, and the selection in this chapter is intended to be a representative selection rather than a full and comprehensive survey.

7.2 Target-specific applications

7.2.1 Nonmetallic mines

A major international problem in countries such as Afghanistan, Cambodia, Angola, Somalia and Kuwait is the presence of large quantities of antipersonnel and antitank mines. Many mines are largely constructed of plastics and the explosive is purely dielectric. These mines are extremely difficult, if not impossible, to detect with conventional metal detectors.

At the International Workshop of Technical Experts on Ordnance Recovery and Disposal in the Framework of International Demining Operations held in Stockholm, Sweden, on 8–10 June 1994, arranged by the Swedish FOA, in co-operation with the United Nations Under-Secretary General for Humanitarian Affairs, the United Nations High Commissioner for Refugees presented the following facts.

Landmines are a humanitarian challenge because they indiscriminately kill and maim civilians. Landmines are weapons that cannot distinguish between a soldier and a civilian, and they remain active for decades. As a result, in the long run most of the victims of mines are innocent men, women and children. Landmines are a humanitarian challenge to UNHCR because they are used in so many conflicts, in such large numbers and so indiscriminately that during war they are a cause of displacement, and after hostilities they endanger the lives of returnees and humanitarian aid workers, delay return and impede reintegration and reconstruction. Landmines are also being used in current conflicts to block humanitarian access.

The scope of the problem can be illustrated by the experience in Afghanistan, one of the countries most seriously affected by mines. According to the Mine Clearance Planning Agency, over a 15 year period an estimated 20 000 civilians have been killed and 400 000 wounded by landmines. The current rate is 4000 killed and another 4000 wounded annually. The UN has been making a dedicated effort to clear the nine million mines. Since 1989, 34 km^2 of land have been cleared of 63 000 mines. The annual cost runs to US$ 15 million to clear 13 km^2. Treatment and rehabilitation averages $5000 per victim.

Afghanistan is not unique. There are an estimated 100 million mines in place in 62 countries. Some of the most affected are countries that are or have been of concern to UNHCR, such as Angola with 9–14 million mines, Cambodia with 8–10 million and Mozambique with over one million. The casualties are correspondingly high. The estimated number of people who have had limbs amputated by mines in Cambodia is 30 000, in Mozambique it is 8000 while in Angola the figure for lower limbs only is 15 000. Worldwide, landmines are taking 800 lives each month, according to the International Committee of the Red Cross.

The large majority of civilian casualties are caused by antipersonnel mines which come in a wide variety of types. Many are designed only to maim. The blast-type antipersonnel mine will cause a traumatic amputation to a foot or leg, often injuring the other leg and genitals as well. Fragmentation mines are far more deadly. Some models shoot hundreds of metal fragments in an arc that reaches out 50 m. Other types spring into the air when triggered and then explode at waist level. Antipersonnel mines can be buried in the ground or placed on the surface and can be set off by pressure, trip wire, remote control or sensors. They can be laid by hand, dropped from airplanes or spread by artillery. Many are made of plastics which means they cannot be located by metal detectors during clean-up operations.

Antivehicle mines are less numerous but more powerful. A mine that can disable a tank will destroy a civilian vehicle and kill its occupants. These mines usually cannot be detonated by a person's body weight alone, although when they are fitted with an antihandling device they become antipersonnel mines. Antivehicle mines are a particular threat to humanitarian aid workers who must travel roads before they have been systematically cleared.

To assist in this problem, surface-penetrating radar has been shown, in the laboratory and in certain well specified environments, to be capable of detecting buried nonmetallic, as well as metallic, mines. The following Sections describe some of the efforts being made to provide a solution to the problem.

The earliest work was carried out by Echard *et al.* (1978) (Georgia Institute of Technology, USA) who carried out a major project on the detection and classification of buried nonmetallic mines using a short-pulse radar. Spectral characteristics corresponding to the physical parameters of the mine targets were observed. Five different frequency-domain classification algorithms were evaluated. These were crosscorrelation, impulse correlation, space partition, nearest-neighbour and Fisher's linear discriminant. The main factor which affected the spectral characteristics of the return signal was found to be the thickness of the mine. In high-loss soils, high false-alarm rates were caused by objects of similar dimensions, for example, rocks and roots, although in low-loss soils the probability of mine detection was 0.9 with a false alarm rate of 0.1. The main conclusion was that the frequency range 0.5–1.5 GHz was sufficient to obtain target-classification information.

Chan *et al.* (1979) (Ohio State University, USA) used a predictor–correlator processor to identify shallow buried mine-like targets. Targets were classified by thresholding out false targets by timing, energy and magnitude. It was found that within a 0.3 m radius from the target centre, the probability of identification was 100%, while the probability of false identification was 1.7%. These probabilities were obtained from an ensemble of 13 mine-like targets and 58 false-target waveforms.

Ilzuka and Freundorfer (1984) found different target signatures from a range of ordnance types detected with a step-frequency radar.

7.2.1.1 Buried mines (FOA, Sweden)
Dr. S. Abrahamson

The impulse radar system used by the FOA in Sweden (Abrahamson *et al.*, 1991; 1992) consists of a radar and an antenna manufactured by ERA Technology, UK. The width of the transmitted pulse is 1 ns with a pulse-repetition frequency of 250 kHz and a peak power of 50 W. The antenna is a crossed dipole optimised for broadband performance (0.2–2 GHz). Data are sampled and preprocessed by means of a digital signal analyser (Tektronix DSA 602) with an amplitude resolution of eight bits, a real-time sampling rate of 2 Gsample/s while the equivalent-time sampling rate can be as high as 1 Tsample/s. The analogue bandwidth of the instrument is 1 GHz which imposes a corresponding restriction on the bandwidth of the complete radar system.

For *in situ* subsurface applications, the entire system was initially mounted on a jeep. The antenna was mounted in front of the vehicle and was scanned mechanically across the track. Data were recorded and the results displayed in real time. Recently, FOA has used a prototype for a man-carried system, where the antenna and a LCD display are carried.

The overall objective of the FOA project was to assess the ability of ground-probing radar to classify detected buried objects. The concept is that targets, depending on their geometrical shape and material content, exhibit different radar-backscattering signatures. Initially, FOA examined basic representations of radar signatures of targets with a minimum of disturbances present by carrying out measurements in an anechoic chamber. FOA found previously (Abrahamson *et al.*, 1991) that a suitable waveform, matching the transmitted pulse, can be designed by digital-filter techniques. An important finding for free-space targets was that useful signature features can be extracted only if a time window is used that is wide enough to accommodate both the initial, specular return and the first creeping-wave return (Strifors *et al.*, 1993).

FOA deliberately initially avoided developing a useful object-classification technique in view of the difficulties connected with the scattering of waves from objects buried in a lossy medium. The digital processing is, however, carried out using a technique similar to that FOA proposes to use for subsurface classification. Field measurements have been directed towards detection of mines, both plastics and metallic, under various soil conditions. For normal soil conditions the detectability seems to be reasonably good.

The first object-classification technique which FOA developed (Gustafsson, 1992) was based on complex poles modelling of electromagnetic scattering, which is related to the singularity-expansion method developed by Baum (1976). The technique has been applied successfully to data from free-space objects, but it needs further clarification when applied to subsurface-object classification.

From the class of time/frequency distributions FOA applied the pseudo-Wigner distribution (Claasen and Mecklenbräuker, 1980a, 1980b, 1980c) to data from free-space objects. Disparate object signatures are obtained in the combined time/frequency domain in cases where spectral features are almost indistinguishable.

A man-carried GPR system for mine detection and classification is currently being developed by FOA and a prototype is shown in Figure 7.1. The display exhibits a depth view of the scanned ground with representations of the amplitude from a subsurface object. A smaller antenna (0.3–3 GHz) and a new

Figure 7.1   Prototype man-carried radar system (courtesy of FOA)

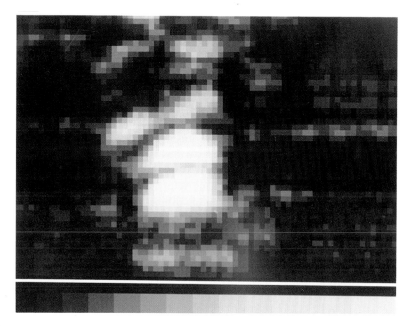

Figure 7.2   Operator display of the radar image of a buried antitank mine (courtesy of FOA)

pulse transmitter (0.3 ns) are the components that have improved system performance considerably. The system performance will be evaluated by simulating real conditions using a sandbox $(4 \times 4 \times 1.5\,m^3)$ in which objects can be buried. The antenna positioning and data recording are computer controlled and so permit automatic data acquisition; a radar image of a mine is shown in Figure 7.2.

7.2.1.2 Nonmetallic mines (ERA Technology, England)

D. J. Daniels

ERA Technology has been involved with surface-penetrating-radar techniques for nearly a decade and was involved with assisting UK MoD (PE) with the first serious attempt to develop a radar to detect buried plastics mines in the Falklands in an active theatre of war in the early 1980s.

Surface-penetrating radar is a promising technical approach to the problem of detecting buried objects such as mines, or investigating the internal structure of nonconducting structures. A high-performance, high-resolution radar system has been used to generate images of the internal composition of many structures. For the requirements for detecting buried mines it is necessary that signals be radiated without distortion over a wideband of frequencies from a suitable antenna. The performance of the antenna is the key factor in determining the overall system performance and has proved to be crucial in determining successful system operation.

In general, antitank mines are significantly easier to detect than antipersonnel mines and experience has shown that it is important to physically scan the antenna in a regular pattern over the mines or use a multi-element antenna.

A plan-image presentation offers the operator the best means of recognising the image generated by scanning over the mines (in a regular pattern) although the importance of accurate registration cannot be emphasized too strongly. The alternative option of a remotely controlled 'crawler' which carries the scanning

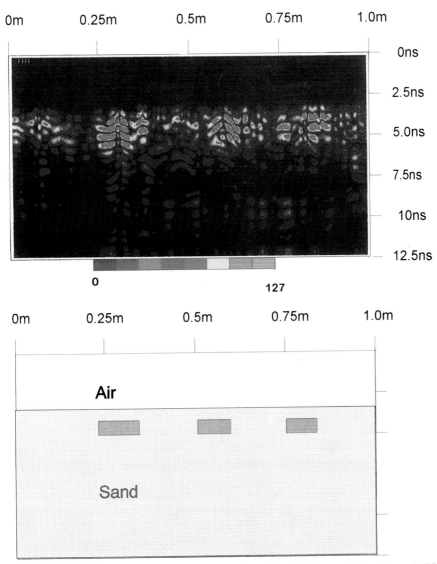

Figure 7.3   Radar image of antipersonnel mines buried in sand (courtesy of ERA Technology)

antenna and radar head should be considered in order to remove the risk of operator fatigue generating irregular scan profiles.

A radar image of antipersonnel mines is shown in Figure 7.3 and was taken with the mines buried in wet sand.

7.2.1.3 Unmanned mine-detection equipment (Daimler Benz Aerospace, Germany)

M. Bartha

'The main tasks of a pioneer troop include the hampering of the movement of enemy troops and the acceleration of the movement of their own troops. In this connection, the mine warfare is of particular importance.' When reading this excerpt from a task description for pioneers the military layman will probably have no idea of the reality of today's mine warfare. In the past decades, the development of mines and fuses has been advanced continuously. Besides the known antitank mines with pressure fuses which may be laid exposed or buried, mines with magnetic fuses, scratch-wire or tilt-rod igniters are now available. Some of these new mine types are free from metal and camouflaged against infra-red-sensors by special coatings. With the progressive development of new mines the chances of detecting or clearing them become increasingly lower. While, for example, in the Second World War the loss of tanks caused by mines was clearly below 30% the US army lost far more than 70% of its tanks in the Vietnam war in minefields and not in the battle.

In spite of great efforts, the development of mine searching and clearing means could not keep up with mine development. The only mine-detecting means currently in service with the Bundeswehr (German Federal Armed Services) are hand-held detectors (ineffective with metal-free mines) and mine probes, very thin metal needles of about 40 cm length by means of which soldiers search the ground centimetre by centimetre for buried mines.

The spectrum of mine-clearing means is wider. The following types are known: line charge (rope or rope ladder with explosive charges), mine plough, mine roller (a heavy iron roller pushed by a mine sweeping tank which mills a ditch of about 40 cm depth and 4.7 m width through a minefield by means of a rotating mechanism of hammers hanging on steel ropes (flails)).

The employment of these clearing means, however, is very time and material consuming, not always reliable and, in the case of mine ploughs and mine clearing tanks, entails much wear and tear depending on the type of ground. Mine-sweeping means can be used effectively only when the position of individual mines or at least of a minefield is known. 'Blind clearing on suspicion' is in contradiction to the high mobility aimed at in the battlefield.

After many years of research and basic development, the German company Eltro developed in 1986 a microwave method which makes it possible to find buried mines up to a depth of about 40 cm. A pulse-radar method via a sensor head emits microwave pulses which penetrate into the ground and a receiver detects the reflected signals which are processed electronically, evaluated, and represented in three dimensions on a monitor. The display shows the boundary surfaces between materials with different relative permittivities. In this way it is possible to estimate the structure of the soil and the form and size of objects (mines, stones). The fundamental capability of this method for mine detection was proved in 1963 by Eltro in the laboratory and on a plane test site.

In July 1984, Dornier was awarded a contract by Bundesamt für Wehrtechnik und Beschaffung BWB after having tendered for the definition of an operational mine-detection equipment (MSG). This operational MSG consists of a control platform and a remotely controlled detection vehicle which, equipped with a microwave sensor by Eltro, is capable of detecting exposed and buried metallic and metal-free AT mines and marking their positions for subsequent clearing.

This equipment is intended to be employed mainly in combination with the mine-clearing tank in order to breach mine barriers quickly, i.e. to clear a passage (4.7 m wide) through a minefield for tanks and operational vehicles. In addition, the MSG detects whole minefields and prepares their clearing.

A lightweight construction using as much glass fibre reinforced plastics as possible is planned for the detection vehicle so as to keep the metal parts below the effective mass for magnetic fuses of antitank mines. This will result in a low total weight whereby, in connection with a tracked drive system, the ground pressure will be low and uniformly distributed. Even if the terrain is unfavourable and the track is only partly in contact with the ground surface a maximum of one third of the initiating pressure is achieved.

To minimise weight, fluorescent, coloured polyurethane foam is used for marking. To avoid confusion between mine and passage markings, different colours are used. The bar-shaped sensor head can follow the form of the terrain at a constant distance by means of hydraulic actuators and an automatic three-axis control. The detection vehicles can in principle roll in any direction. It can move and search with the sensor head in front or vice versa. However, the sensor head should follow behind. In this case the fragment-protected vehicle will shield the sensor head against antipersonnel (AP) mines which can already be initiated in front of the vehicle by the 'rake' of glass-fibre-reinforced plastics bars.

For remote control and observation of the sensor head, the detection vehicle is equipped with three television cameras. The sensor-head observation camera is located at the bottom of the vehicle and trained to the lower part of the sensor head and the terrain below. Each of the two other nonadjustable cameras will look in one of the two possible driving directions.

The video images and sensor data are transmitted to the control platform via single optical fibre. The control platform transmits its control signals to the detection vehicle via the same optical fibre in the opposite direction. For an emergency recovery, in case the optical fibre is damaged, a narrow-band radio channel will be available.

Control of the mine-detection equipment requires two operator places in the control platform, one for the remote control of the detection vehicle, the other for evaluating the sensor signals in the three-dimensional-monitor.

Figure 7.4   Unmanned mine detection equipment (courtesy of Dornier)

To verify the design of the operational mine-detection equipment and mainly to investigate the microwave sensor in the field, a proof-of-principle prototype was built. By means of this prototype all critical functions and components were realistically tested.

The proof-of-principle prototype was completed in August 1986 and subjected to a comprehensive trial at Dornier. The test results confirmed the function of the microwave sensor in the field and the basic concept of the operational mine-detection equipment. Further tests over six months with the prototype were conducted by the test agency 52 of the Bundeswehr in Oberjettenberg. A photograph of the vehicle is shown in Figure 7.4.

7.2.2 Pipes and cables (British Gas Research and Technology, UK)
H. F. Scott

British Gas has investigated the use of surface-penetrating radar as a means of detecting and mapping buried plant, both metallic and nonmetallic, and at the Engineering Research Station has pursued the development of the prototype to the stage where six evaluation units were manufactured.

The best performance from a surface-penetrating-radar system, particularly one for the detection, at high resolution, of shallow objects, is obtained when the whole of the system is designed around a specific target type or geometry.

The detection of utilities' plant imposes a particular set of constraints on the design of an effective surface-penetrating-radar system. The majority of buried plant is within 1.5–2 metres of the ground surface, but it may have a wide variation in its size, may be metallic or nonmetallic, may be in close proximity to other plant and may be buried in a wide range of soil types involving large differences in electromagnetic absorption. As a result, obtaining adequate penetration of the emitted radiation, coupled with good resolution of neighbouring plant, is not straightforward and some compromise has to be accepted in the design. The particular feature of utilities' plant which may be used to advantage in the design of a radar detection system is that almost all the objects sought are long and thin, which results in a particular action on the polarisation of a reflected wave. This feature has been used in the design of the British Gas pipe-locating radar, described in more detail below.

Ground-probing radar equipment, being an emitter of radio signals, is subject to licensing and control in the country of use. An additional requirement, therefore, is that any equipment used must comply with the legal requirements locally in force, which in the UK are set by the Department of Trade & Industry. It is also essential that the radiation levels are below the relevant safety thresholds.

A pipe-location system should have the ability to detect and map all classes of buried pipe and cables. Although not essential, the further attribute of being able to detect other subsurface features such as voids and ground obstructions (concrete etc.) would be an advantage.

The ability to detect all buried plant down to a depth of 1.5 m is desirable. If this could be achieved, approximately 90% of all buried plant could be located under all conditions. In practice, there is an approximate relationship between pipe diameter and burial depth; for example, 25 mm polyethylene gas service pipes are not often laid deeper than 500 mm cover, while plant laid at more than 1 m cover is predominantly of diameter 125 mm or greater. Consequently, the penetration characteristics of a practical pipe and cable locator must be consistent with the observed depth distribution of diameters.

All plant-location equipment is limited in its ability to resolve closely spaced objects. For example, the horizontal resolution of conventional inductive locators is usually expressed as a function of burial depth: they can resolve adjacent plant when their spacing, in the horizontal plane, is approximately the same as their depth. A desirable resolution is the ability to distinguish multiple buried plant as separate targets when they are spaced more than 150 mm apart, in either depth or plan.

In the British Gas radar-system equipment, the antennas are planar logarithmic spirals, interleaved so as to be electrically orthogonal. The impulse is approximately 1 ns in duration and the antenna output is in the frequency range 200–1000 MHz. The use of a planar spiral antenna has three principal advantages. First, very broadband transmission is possible so that there is no need to select different antennas for different circumstances. Secondly, the emitted radiation is circularly polarised so that a pipe will be detected at

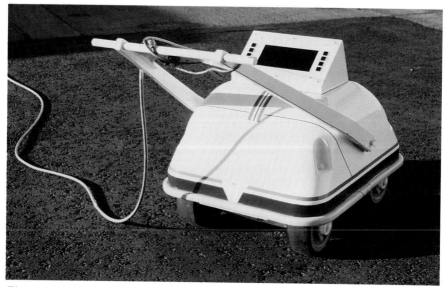

Figure 7.5   Radar head (courtesy of British Gas)

whatever horizontal angle it may lie relative to the antenna. Third, a planar design means that it may be positioned close to the ground surface while still maintaining mobility for rapid surveying.

A rotating-antenna system has been found to give a significant reduction in levels of ground clutter and noise, when the received signals are suitably processed. Data are collected during each half or full rotation of the antenna as the trolley on which it is mounted is moved forward over the ground. Each set of data, after processing, is stored along with an indication of the position along the surveyed line to which it corresponds. The position is provided by a shaft encoder attached to one of the trolley wheels.

The radar system consists of two main parts. There is a mobile trolley, shown in Figure 7.5, and a base station which resides in the support vehicle, shown in Figure 7.6. The two are connected by an umbilical cable to provide power and to transfer data.

Primary power is supplied by a portable petrol generator of approximately 1 kW capacity.

The trolley contains the impulse generator which is connected through a rotating microwave joint to the transmitting antenna. Together with the receiving antenna, this is contained in a cylindrical drum rotated by a drive motor. The received signal is transferred to an amplifier and sampling receiver which delivers data, digitised to 16 bits, to a real-time digital signal-processor unit. Processed data are taken by the umbilical link to the base station for storage. The link also provides power to the antenna-drive motor and to a voltage-converter unit which supplies the electronic modules.

Mechanically, the trolley chassis plate is mounted on a plastic suspension

Figure 7.6   Radar base station (courtesy of British Gas)

frame, with leaf-spring and wishbone damping, supported by four plastic and rubber wheels. To one of these wheels is attached a shaft encoder linked to the processor unit by optical fibres. There is a glass-reinforced plastic cover, in the top of which is fitted a back-lit liquid-crystal display to provide essential information to the operator.

The base section is a custom-built computer unit using a versatile modular system of construction. It contains a hard disk for data storage, a 68020 microprocessor with a floating-point coprocessor, a framestore to drive the VDU and a tape streamer for archiving. Information may be entered via a keyboard, but operation is otherwise controlled by a mouse or by a touch screen. The base station also contains AC-to-DC power converters for the whole system.

The radar emissions are in the frequency range 200–1000 MHz, at a mean power of 10 mW. The transmitter has been Type Approved by the UK Department of Trade and Industry, and it is operated under a telemetry licence.

Maximum benefit is obtained from a survey of buried plant when the output is available in the form of a map. To that end, the area to be surveyed is scanned by moving the equipment over a grid with a rectangular boundary, with each stored data item being logged with the co-ordinates of the point where it was collected. By means of trigger signals from the encoder attached to one of the trolley wheels, data are collected every 0.1 m of linear travel as the trolley is propelled along a straight line, usually defined by a stretched cord near the ground surface.

The present umbilical cable allows lines up to 100 m in length to be surveyed. The spacing between adjacent lines of the grid is determined according to the circumstances of the work to be performed and is affected by any prior knowledge of the plant contained in the area under investigation. An extreme case is when the lines are set at 0.1 m spacing to give a square grid. It is often convenient to subdivide large areas into a number of smaller ones for survey purposes.

Speed of survey is normally set at one linear metre per second, or an area rate (not allowing for turnaround time) of 6 m^2 per minute at the maximum data density; an increase in data quality can be achieved with a reduction to 0.5 metres per second. The data can be stored in the base station at the same rate as they are collected and are available for viewing in a pseudocolour representation once the survey of a grid has been completed.

At the beginning of a survey the operator enters at the base station a range of parameters, such as grid size and location, as well as notes relating to ground conditions and weather. During the data-collection stage the base station performs the data storage with no operator intervention, but at the end of the survey the results are available for examination on a VDU in a range of formats, and some image-processing techniques may be applied, if necessary, to aid interpretation. Data may be archived onto magnetic tape for attention at a later time, if required.

It is usual to have two operators working a site, with the equipment transported in, and operated from, an estate car. In addition to the operation of the radar equipment, it is necessary to perform some basic land surveying to establish the grid position in relation to the site as a whole and to record this on a map or plan. A simple CAD package is used to generate such plans (Gunton and Scott, 1987).

Presentation of results

After a grid survey has been performed, the data stored in the computer are in the form of a three-dimensional block. To examine this block, for indications of linear targets, it is possible to section it in either vertical or horizontal planes. The data contained within one plane are normally, represented in pseudocolour, that is, a different colour, or shade of colour, is assigned to a data point depending upon its numerical value.

An example of a field trial result is shown in Figure 7.7. This is a horizontal section of a survey of part of a test site. At a depth corresponding to this section, there are visible five items of plant. Other items are apparent at other depths. All items of plant detected in the survey were later confirmed as those known to be present in the test area. The soil was light clay, covered with loam.

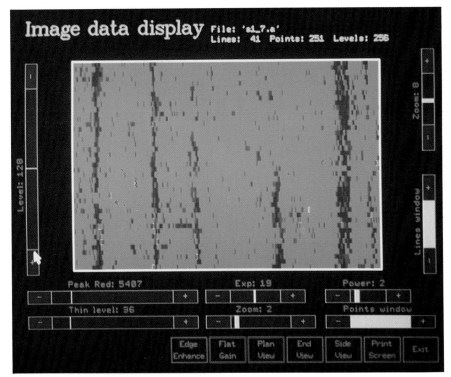

Figure 7.7   Radar image of buried pipes (courtesy of British Gas)

7.2.3 Internal inspection of pipes (ERA Technology, UK)

D. J. Daniels

Much of Britain's 475 000 km network of sewers and water mains is continuously assessed for structural integrity. Many of the sewers in major cities are over 150 years old, dating back to the start of the industrial revolution. The replaceable assets of the water system were worth, at the beginning of the 1980s, over £50 billion (80% of which is accounted for by the underground sewers and water mains). Sewer collapses can cause dangerous subsidence of road surfaces and sometimes nearby buildings.

One of the problems faced by the water authorities is a lack of detailed knowledge of what there is underground. Teams trying to stabilise collapsed sewers often find uncharted ones underneath. However, the extensive use of television cameras and a radar to detect voids underground is helping to make the unpleasant task of surveying much easier.

The proper condition of sewers and pipelines is vital to ensure that dirty water is transported without pipeline leaks causing contamination and possible health hazards. It is generally difficult to establish the condition of the material surrounding the pipe wall. Radar systems can be used from inside the pipe to

Figure 7.8   Pipe-inspection antenna (courtesy of ERA Technology)

assess the state of the surrounding ground. The pipe should, of course, be constructed from nonmetallic material.

ERA Technology has been involved in the development of a purpose-designed radar system to work from inside the sewer to produce a radar image of the material adjacent to the sewer wall. The radar antenna consists of four quadrants which enable four sectors of the sewer to be surveyed. The antenna has a diameter of 180 mm and is designed for use in sewer pipes of an internal diameter of 200 mm. Several trials have been carried out with the antenna and results show that small voids of 50 mm diameter can just be detected. A photograph of the radar antenna is shown in Figure 7.8 and a typical test result in Figure 7.9. Within the test rig and effectively on top of the 200 mm diameter pipe under test are a number of artificial voids. Their reflections can also be seen in the radar image. The radar reflection from the 150 mm diameter pipe, which crosses the pipe under test, can be seen on the righthand side and then the reflections from the 200 mm void and the 100 mm void can be seen adjacent to the pipe under test. Also to be seen is the reflection from the interface between the top soil and the air which appears inverted in the radar image.

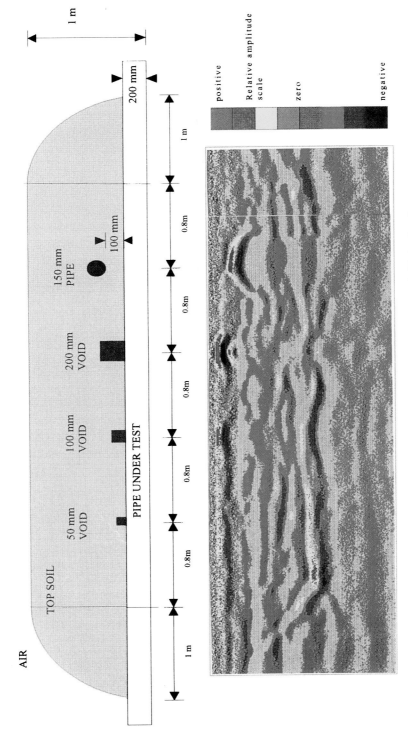

Figure 7.9 Radar image from one channel of a pipe inspection radar (courtesy of ERA Technology)

7.3 Nondestructive testing

7.3.1 Concrete (University of Liverpool, UK)

Prof. J. Bungey

Radar testing of structural concrete is part of a research programme at Liverpool University that has been undertaken since 1990 by a team led by Prof. J. H. Bungey. The work has been supported by SERC/EPSRC in co-operation with several industrial organisations as well as Prof. Forde of Edinburgh University (Bungey and Millard, 1993; Bungey *et al*, 1991, 1993a, 1993b; Millard *et al.*, 1993; Shaw *et al.*, 1993). Attention has been concentrated on assessing the capabilities of commercially available apparatus applied to practical problems in the laboratory and on site. This has involved specially cast concrete-slab specimens as well as the use of an oil–water emulsion-tank facility developed to simulate slabs with a range of concrete properties, and to permit rapid variations to parameters such as reinforcing bar, prestressing duct or void configurations. This facility avoids the need for casting and curing of large numbers of concrete specimens and has permitted an extensive library of characteristic responses to be established.

Measurements were carried out using centre frequencies between 500 MHz and 1 GHz by means of GSSI SIR 8 and SIR 10 equipment.

A coaxial transmission-line system has been developed to assess the relative permittivity and electrical conductivity of both fluids and hardened-concrete specimens over a wide frequency range, and results are shown in Figure 7.10. This has assisted development of emulsions for the simulation tank and permits concrete specimens to be removed for conditioning. It has been used to assess the effects of concrete moisture content, which is a key factor in the successful use and interpretation of radar surveys. Work is continuing to establish fundamental properties for a wide range of concrete types using this facility which has been refined to improve capabilities.

Computer modelling, using specially developed two- and three-dimensional ray-tracing programs and a commercially available finite-element package has been actively progressed and finite difference techniques are under investigation.

The results from the studies at Liverpool University have permitted realistic assessment of the capabilities and shortcomings of the technique at its current stage of development with commercially available apparatus and software. The effects of reinforcing steel in masking deeper features are of particular importance, and effective minimum limiting spacings of between 100 mm and 200 mm have been suggested for 1 GHz apparatus, depending upon bar diameter and depth. Similarly, it may be possible to identify individual reinforcing bars at spacings down to 100 mm centres depending on bar size and cover. Void and crack detection are influenced by shape, size, depth and whether they are air or water-filled. Cubic voids are easier to detect than spherical, and water filled voids easier than air-filled. Whilst 50 mm air voids

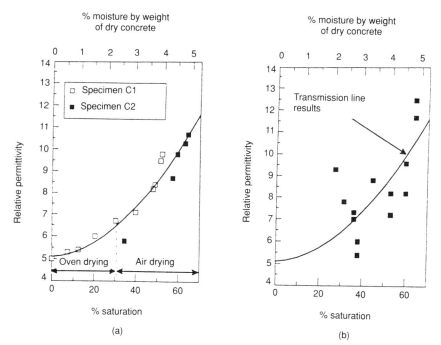

Figure 7.10 *Measured variation in the relative permittivity of concrete as a function of moisture content (courtesy of University of Liverpool)*
a Transmission line specimens
b Slab specimens

could generally be detected, smaller sizes down to 20 mm required very precise apparatus adjustment, highlighting the practical difficulties in locating a small void or similar feature whose presence is unknown. Buried honeycombing is particularly difficult to detect, and highlights the critical aspect of optimum transducer orientation to minimise reinforcement influences.

It is clear that further improvements in interpretation capability hinge on knowledge of the fundamental electrical properties of concretes, with especial reference to the influence of pore fluids. For example, accuracy of current depth estimates to reflective features may range from about +45% to −15% owing to such uncertainties with a commonly adopted relative permittivity for concrete of 6.5, although this can be improved by calibration drilling.

Signal processing can be most valuable when trying to detect or identify small or complex features, but is expensive and it is questionable whether extensive use of such facilities is justified in straightforward applications. Simple equipment giving basic monochrome plots may similarly often be adequate.

Ray-tracing programs have been very successful in simulating radar responses to complex configurations. Finite-element modelling incorporating open boundary elements is also highly successful in demonstrating the propagation of radar waves in materials with varying properties and discontinuities, and simulated scan patterns can be generated which closely match those obtained in practice. Such results provide a valuable aid to understanding and interpretation.

7.3.2 Buildings (Building Research Establishment, UK)

Dr. R. de Vekey

Radar will give reflections for many of the boundaries within built structures and as a result can give information on (i) presence/position/size of nonvisible (hidden) features of buildings and (ii) on the quality of known features. For the scale of objects in built structures the bandwidth 0.3–1 GHz is the most appropriate.

The important boundaries in built systems are:

(*a*) at the interface between solid material and a significant void;

(*b*) at the interface between good concrete and poor concrete/mortar;

(*c*) at the interface between concrete/masonry and insulation materials;

(*d*) at the interface between large metal inclusions, e.g. reinforcement and other components;

(*e*) at the interface between dry material and damp/wet material (since moisture alters the relative permittivity);

(*f*) where signals are reradiated from metal objects which tune in some way to the wavelength used. Small objects such as ties, bolts etc. are likely to respond in this manner to the wavelength used for these investigations.

The main problems are:

(i) If the material is wholly saturated or very wet, the signal attenuation is such as to prevent the satisfactory use of the equipment.

(ii) New damp concrete or mortar has a very high attenuation and is virtually impossible to investigate.

(iii) Continuous metal, i.e. sheet or fine mesh reinforcement, will reflect all the incident ray thus preventing any investigation of what lies behind it.

(iv) Some components of buildings are so complex that it is difficult to resolve the signals from the various reflecting objects.

7.3.2.1 Applications

Most of the work carried out by the Building Research Establishment was carried out using a GSSI radar system on the following artefacts: walls and other structures made from mortared or dry-stacked bricks, blocks, dressed stone blocks, rubble, plain and reinforced concrete.

Masonry

Although masonry is complicated by the array of bricks and joints, it is often relatively simple in structure, being composed of layers or sheets joined by bonding or by metal ties.

If a wavelength or antenna orientation is chosen so that it is optimum for the dimensions of the required target then the following features can be detected in solid masonry; the position and layout of hidden voids such as chimneys, air ducts and other features, internal cracking, the thickness of walls, infilling of walls with rubble or poor concrete, the numbers of wythes (or rings in arches) the presence of buried metal such as cramps, reinforcement, fixings, straps, lintels, hidden pipe runs, chases and electric cabling and, where localised, the position of damp areas. Work on detection of hidden objects is reported by Botros *et al.* (1984) and Carr *et al.* (1986). Some work on masonry tunnels has been reported by Transbarger (1985).

In addition the following features can be detected in cavity walls: the thickness of the leaves, the cavity width, the layout and, to some extent, type of wall-ties, the presence of larger objects or blockages of the cavity. Only very limited research has been carried out and very little has been published. The technique has been covered in a review by de Vekey (1988) and some examples of applications are given by Baston-Pitt (1992).

Concrete system walls and floors
Typically concrete walling and flooring elements consist of planes (flanges) of reinforced or pre-stressed concrete planks or prisms connected at intervals by webs and incorporating voids. Often site concreted joints and additional reinforcement are also present.

Walls may additionally have a finish such as wet plaster or dry lining of plasterboard over bats. The hollow sections may be filled with sand to increase the sound insulation or with thermal insulation materials. Floors typically have a finish on the soffit which may incorporate nominal reinforcement and a screed on the top surface which may incorporate heating tapes, reinforcement, pipes, conduit etc. They may also have additional reinforcement in the site-cast joints. Externally exposed components may also incorporate layers of insulation and surface weather protection.

Radar can give information on the dimensions, position, voids etc., in the various layers. A brief description of an investigation of some system flats using Radar is given by de Vekey *et al.* (1989, 1992, 1993).

Joints in concrete system buildings
These joints consist of a complicated junction between the top of a reinforced wall slab, in intersecting floor and the base of the next slab. They incorporate site-placed tying reinforcement, concrete and void-filling dry-pack mortar and insulation layers and a facing panel on the exterior. Although this represents the most difficult three-dimensional structure to interpret, it has probably been the largest commercial application to date. On real structures it is very difficult to interpret the data from normal analogue radar because of the complexity of the structure and the number of reflecting features. Analysis of the investigation of dry-pack quality on Ronan Point gave a disappointing correlation between radar data and visual exposure of the beds. De Vekey *et al.* (1989) obtained some

more statistically valid information by investigating model joints in a laboratory trial.

7.4 Geophysical applications

7.4.1 Peatland investigations (Geological Survey of Finland)
Dr. P. Hanninen

The Geological Survey of Finland (Dr. P. Hanninen) has carried out hundreds of surveys of peatland. Peat thickness results obtained by means of ground radar in peatland investigations are substantially more comprehensive than those yielded by drilling. The radar profile provides data on the thickness of peat layer and the nature of the underlying mineral soil (Figure 7.11) as well as information on the internal structure of the peat (Figure 7.12). Accurate depth maps for the bog concerned can be provided by combining radar cross-sections. Although it is very difficult to identify layers of gyttja, e.g. lake mud, which are rich in organic material, they can be distinguished from peat fairly easily when they are rich in mineral material.

Ground-penetrating radar can be utilised best for the examination of open, unditched peatland areas, but is more difficult to use in bogs which support forest or contain a dense network of ditches. The results can be used for the planning of ditching (outlet ditches, environmental ditches, field ditches), road networks and precipitation basins in peat-mining areas, and for evaluating the quantities of garden peat and fuel peat obtainable from peat-mining areas.

80–500 MHz antennas can be used in peatland investigations, the low-frequency types lending themselves best to the measurement of peat-layer thickness and the high-frequency types to obtaining data on the surface layer

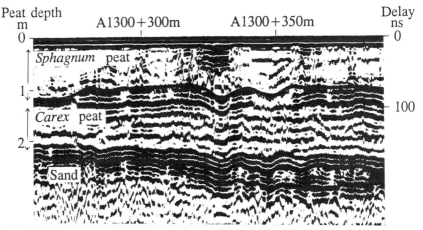

Figure 7.11   Radar cross-section of a peat bed (courtesy of Geological Survey of Finland)

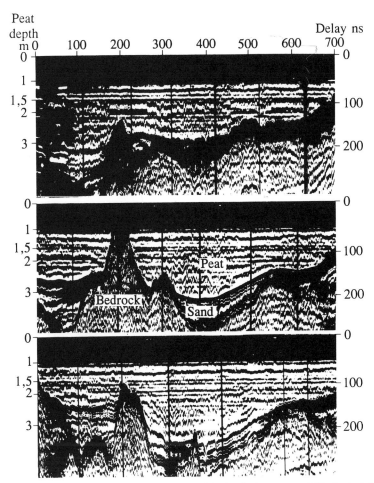

Figure 7.12   Radar cross-section of peat and bedrock (courtesy of Geological Survey of Finland)

and on moisture differences between peats. The use of high-frequency antennas is hampered by their sensitivity to interference, as a result of which there must be no other medium between the antenna and the object measured. This means in practice that the antenna cannot be pulled on a plastics or glass fibre sledge, for example, but must be moved along the bog surface at walking pace.

7.4.1.1 Soil contamination (Delft Geotechnics, Netherlands)

Dr. J. K. van Deen

Spillage of polluting substances heavier than water is one of the serious problems of environmental engineering. From the source of pollution the substance

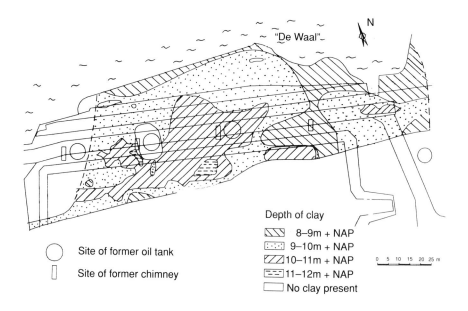

Figure 7.13   Contour map of reclaimed site (courtesy of Delft Geotechnics)

migrates more or less vertically through permeable layers until it meets an impervious (e.g. clay) layer. From there on it spreads horizontally, collects in shallow depressions of the clay layer, follows the slope of the clay boundary and may even leak through missing parts (holes) of the clay layer. It is generally difficult to discern the pollutant itself directly in the radar reflection; however, delineating accurately the upper surface of the clay layer is as effective, since it guides the sampling strategy to the depressions where large amounts of pollutant may be expected and to the holes with their possible risk of leakage to deeper layers. Figure 7.13 shows a contour map of the top of the clay layer in a situation where the primary surface (clay) had been artificially covered by a 3–5 m thick sand layer as a foundation layer and working area for industrial activities. The Figure shows clearly the (former) topography as well as a large missing part in the clay at the west side of the terrain.

7.4.2 Geological structures (TNO Institute of Applied Geoscience, Netherlands)

Dr van Overmeeren

The accuracy of present surface-based geophysical methods for the study of shallow subsurface geology, i.e. down to 40 m below the surface, is not always adequate. A higher definition is required, for example, for dealing with soil and

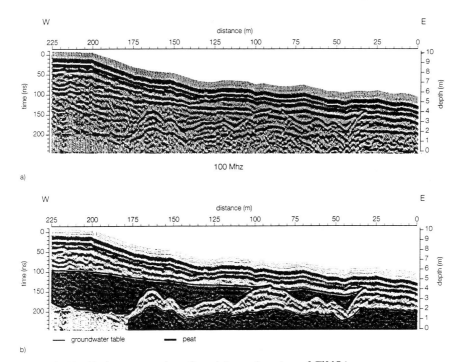

a)

100 Mhz

b)

——— groundwater table   —— peat

Figure 7.14   Radar cross-section of sand dunes (courtesy of TNO)

groundwater pollution, for measuring the groundwater level, for the construction of deep foundations and tunnels etc. A georadar system has been developed at TNO Institute of Applied Geoscience (Dr van Overmeeren) and combines a greater exploration depth with a more complete integration of acquisition, processing and interpretation of data, while remaining highly cost-effective. The georadar system will contribute to the more efficient management, exploitation and protection of the subsurface environment.

High-resolution seismic-reflection methods fail to provide subsurface images of the shallow zones (< 40 m) which become increasingly more important in hydrogeological and, often related, environmental investigations. Ground-penetrating radar (GPR) at low frequencies (< 200 MHz) is a geophysical technique that is similar and complementary to seismic reflection methods; based on wave propagation, highly detailed continuous subsurface images (sections), showing reflecting interfaces between contrasting layers, are obtained by both methods. The deeper zones where seismic reflection reigns are out of reach for GPR. In the shallow zones, however, GPR is indispensable for obtaining images of high resolution.

For several decades, GPR has been used for engineering surveys, at depths of several metres below the surface. Only recently, however, with the advent of digital instrumentation and low-frequency antennas, have exploration depths been extended (in favourable terrain) to several tens of metres, opening the door for GPR to the groundwater domain, hence the name 'georadar'.

Georadar for groundwater is particularly applicable to sandy sedimentary environments of high electrical resistivity.

In the Netherlands its use is confined to the elevated areas of the east where sandy deposits in push moraines and terraces prevail, and groundwater is found at depths ranging from 10 to over 40 m. In the lowlands of the western part of the country, clayey deposits and brackish and saline groundwater at small depths having low electrical resistivities mostly impede the use of georadar because of attenuation of the radar waves. The coastal sand dunes here constitute radar-friendly exceptions and typical radar cross-sections are shown in Figure 7.14.

7.4.3 Soil erosion (Delft Geotechnics, Netherlands)

Dr. J. K. van Deen

Part of the coastline of the Netherlands consists of dykes, which are at many places provided with a revetment of rip-rap, asphalt or concrete blocks to protect the clay surface layer of the dyke against wave attack and erosion. Behind the concrete blocks, however, erosion channels develop through erosion

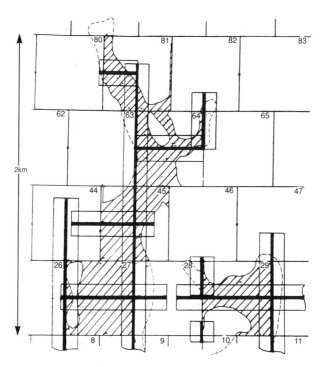

Figure 7.15   Radar map of erosion channels (courtesy of Delft Geotechnics)

of the clay by water entering through the seams at high tide and flowing behind the blocks. Owing to the clamping of the blocks, these erosion channels are generally not visible from the outside but may nevertheless develop into extended and dangerous proportions. Under large wave attacks the revetment may instantaneously collapse leaving the soil unprotected to the waves.

To determine the presence and extent of erosion channels, measurements were performed along traverses in a square grid of 50 × 50 cm. After the data had been analysed and interpreted in terms of erosion channels, part of the 20 cm-thick blocks were removed and an accurate record taken of the features showing up. Figure 7.15 shows a map of the erosion channels as recorded below the blocks removed, in conjunction with the data records of the particular radar sections. Quantitative comparison of 'radar' and 'real' holes demonstrated a success percentage of 85% of the scan line length. Most of the remaining 15% was located directly beside the channels. The measurements are nowadays repeated yearly in order to follow the evolution of the erosion patterns and to check the effectiveness of remedial measures such as grouting.

7.4.4 Coal and salt

Surface-penetrating radar has been usefully used in the exploration of rocks and minerals. The earliest work was carried out by Cook (1975), Unterberger (1978) and Unterberger (1985) and coal, rock salt, oil shales and gypsums as well as limestones and granites have been investigated. Cook (1975) carried out numerous experiments and developed a predictive method for determining the depth of probing as a function of material loss in decibels per metre per gigahertz. Cook's paper provides details of the dielectric properties of 38 different types of rock, and a selection is reproduced in Table 7.1 of the loss measured at 100 MHz and the maximum range for a radar with a 100 dB dynamic range.

Unterberger (USA) carried out a number of investigations into the use of radar to probe salt structures. Using a 230 MHz pulse radar he investigated the Pine Prairie salt dome in Louisiana. An interpretative cross-section and a contour map of the near flank of the dome were confirmed by surface-gravity data. The lower flank position was confirmed by slant drilling, which intercepted the salt sediment interface at 2700 m. A similar radar was used in the Cote Blanche salt dome and a one-way penetration of 900 m was achieved. An included pillar discontinuity within the salt seam was discovered.

An interesting technique for measuring the thickness of rock and coal by means of a noncontacting sensor has been developed by Chufo (1992). The technique used a sector network analyser and IEEE 488 bus controller and a servo-controlled L-Band antenna positioner.

The antenna was moved so as to vary the distance between the material and the antenna over a distance of 40 cm in increments of 1.27 cm. At each antenna position the reflection coefficient of the material was measured over a frequency range of 0.6–1.4 GHz at 2 MHz intervals.

Table 7.1   Cook's tables

Material	Country	Loss	Detection range
		dB/m	m
Granite, (dry)	Switzerland	2.7	57
Granite, (wet)	Switzerland	4.35	40
Limestone	New Mexico	6.9	28.3
Limestone	Texas	7.03	27.74
Limestone	Italy	7.35	27.13
Coal	Pittsburg	8.86	21.95
Coal	Virginia	12	20.11
Concrete	Various	12.8	18.89
Coal	Colorado	14.2	17.37
Oil shale (rich)	Colorado	15.0	16.13
Tar sand (rich)	Alberta	15.5	15.54
Gypsum	England	15.8	15.24
Quartzite	New Mexico	17.6	14.63
Schist	Washington DC	18.5	14.02
Concrete (wet)	Texas	19.6	13.72
Limestone (wet)	Arizona	21.7	13.41
Limestone (wet)	Italy	25	10.97
Limestone (wet)	New Mexico	22	13.1
Sandstone	New Mexico	22.4	12.8
Gypsum	England	23.5	11.88
Coal	Ohio	28.3	10.06
Schist (wet)	Washington DC	29.5	9.75

The principle of the technique uses a spatial-domain concept to modulate the received signal so as to remove reflections that are not along the axis of the antenna motion.

The measurement system is calibrated by measuring a reflecting metal surface. The theory of the calibration method is based on a revised version of the linear reduction method and the technique has been used to measure the reflections from the coal/shale or coal/rock interface in order to determine the coal thickness remaining between the cutter head of a mining machine and the mine roof rock above the coal.

The technique can also be used to measure the relative permittivity of each layer of a multilayer medium such as a coal seam.

Chufo (1992) showed that the technique could be used to measure a variety of materials such as salt, granite and coal with accuracies generally within 5%. A typical measurement is shown in Figure 7.16.

A number of ultrawideband impulse radars have been investigated for coal-seam measurements, in the USA, South Africa and the UK. Daniels (1980), on

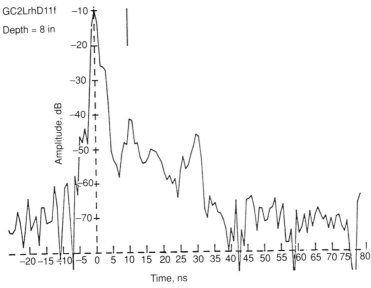

Figure 7.16   Radar measurement of coal thickness (courtesy of US Bureau of Mines)

behalf of the then National Coal Board (UK), showed that measurement of coal-seam thickness of up to 0.2 m was feasible and inclusion of layers of iron pyrites could be detected at thicknesses up to 0.3 m.

7.4.5 Rocks (CEA, France)

Dr. S. Tillard

As part of a project on the qualification studies carried out on nuclear-waste disposal in deep geological formations, part of the R&D work of the Waste Storage and Disposal Department of the Atomic Energy Commission (CEA) in France (S. Tillard) is focused on the optimisation of geophysical techniques. The objective is to locate geological discontinuities likely to allow the migration of radionuclides from the disposal chambers to the biosphere. The acquisition of the parameters needed for other modelling of the hydro-mechanical behaviour of geological rock formations is the aim of the programme.

The core of the CEA research is the evaluation of continuous and non-destructive investigation methods either from boreholes or in galleries in the vicinity of disposal. The principle of three-dimensional visualisation of the massif geology using two-dimensional soundings over the largest possible range of investigation distances was chosen. The objective of an optimised probing range that is compatible with the dimensions of the geological discontinuities is dictated by a concern to minimise prospecting costs and also to avoid any excessive disturbance or weakening of the natural environment. A borehole, even refilled, is liable to constitute a preferential pollution pathway during storage lifetime.

For disposal in a crystalline formation, efforts in geophysical research must be focused on locating and characterising hydraulically effective fractures. The radar technique appeared a promising technique. Granite is a mechanically dense and electrically resistant material, with a low propagation attenuation coefficient for electromagnetic waves. The latter are particularly sensitive to the presence of water. The work undertaken at the CEA has consisted in studying the suitability of radar tools. On account of, on the one hand, the similarity of the acquisition, restitution and interpretation principles of radar data to those for seismic data, and on the other hand, the importance of signal processing in seismic analyses, the ultimate objective is to work on the transposition or adaptation of programmes developed by seismologists to radar records.

This task implies, first of all, fully mastering the operating conditions of a georadar and defining the physical processes which determine the propagation of a high-frequency electromagnetic wave in a geological environment. Because of the cost of drilling or of a borehole measurement campaign, it was deemed more economic to conduct preliminary studies with an adaptable surface equipment (EKKO IV Pulse manufactured by Sensors & Software).

One of the problems treated was the determination of variation of wave-propagation velocity with depth. This subject has seldom been dealt with in the literature. However, a knowledge of the variations of this parameter as a function of the propagation distance is essential for working with seismic migration software and defining geological sections graduated in distance rather than in two-way travel time. Moreover, the accuracy with which this value can be known allows an assessment to be made of the uncertainties affecting radar location of geological discontinuities or anomalies. In many cases, radar operators content themselves with extrapolating the direct underground velocity to the whole formation. The applicability of this value for the whole structure is debatable, especially in the case of superficial alteration of the terrain. Depending on the configuration of the environment of interest, the CEA research team focused its attention on the portion of terrain affected by direct propagation. It established that a road surface of about 25 cm over dry sand or a concrete surface of the same thickness covering the ground of a gallery dug in granite did not have a quantifiable effect: taking into account the uncertainty that affects any velocity calculation, the velocity obtained with the transmitter/receiver in contact with the geological material was similar to the value found in the presence of an interface.

The approximation mentioned above may not be sufficient, in particular for a deep investigation. A study was conducted in order to evaluate the relevance of velocity analyses as they are programmed in seismic studies based on 'normal move out' corrections (determination of root-mean-square velocities and interval velocities). The site chosen for this study was a granite quarry made of a succession of subhorizontal banks (stratiform rock). The radar experimentation aimed to provide a three-dimensional visualisation of the limits of these banks, similar to millimetre fissures (see Figure 7.17).

The interpretation of the RMS data relative to the 'wide-angle' or 'common-midpoint' (CMP) sections, recorded in different directions for apparent different dips and using three transmission frequencies (50, 100 and 200 MHz), made it possible first to verify that no velocity dispersion occurred with frequency and secondly to show the influence of slight dips ($< 15°$) that we had considered disregarding at the beginning of the study. As a result, only the data recorded according to the CMP configuration are now utilised.

Despite the care taken in determining the RMS velocities and in reading the arrival times of each reflection on a great number of high-quality records, the instability of the interval velocities calculated using Dix's formula led us to reject this technique for future work. It seemed more satisfactory to limit ourselves to defining the propagation law from the RMS profile, reduced by a few percent as is the practice in seismic analyses. The CEA research team showed the relevance of a velocity analysis made in this manner by comparing the depth scale obtained by processing with the reflector depths read on cores from an available borehole in the quarry and with the depths calculated with recourse to the direct wave velocity or bibliographical values. However, these analysis methods borrowed from seismic research have limits inherent in, among other things, the hypotheses set for simplifying the equations describing wave propagation. For sites with high reflectivity such as our test quarry (fissure density = 1.3 fissure per metre), it seemed difficult to determine a velocity law with an uncertainty of less than 10%.

The purpose of velocity calculation is not solely for locating deep discontinuities; it was also shown that, when direct propagation velocities are determined with low uncertainties, these values can be utilised to differentiate rocks of the same nature but of different composition in the same site (a numerical example: the differentiation of an oolithic limestone from a comblanchien limestone by means of wave propagation evaluated at 77 ± 4 m/μs for the first compared with 85 ± 4 m/μs for the second). In this case, the inversion of the velocities in dielectric permittivities allows a qualitative interpretation of the radar signal to be made in petrophysical parameters (porosity, clay content etc.).

Large velocity variations during propagation in an anisotropic environment have been observed during radar prospection in schists. For propagation parallel to the direction of the schistosity, a velocity twice as high as that calculated when the propagation is perpendicular to the same schistosity was found ($\sim 100 \pm 5$ m/μs and $\sim 50 \pm 10$ m/μs). The uncertainty affecting these velocities was higher for perpendicular propagation because of the greater attenuation of the waves that occur for an electric field directed parallel to the schist layers. These results were confirmed by dielectric measurements carried

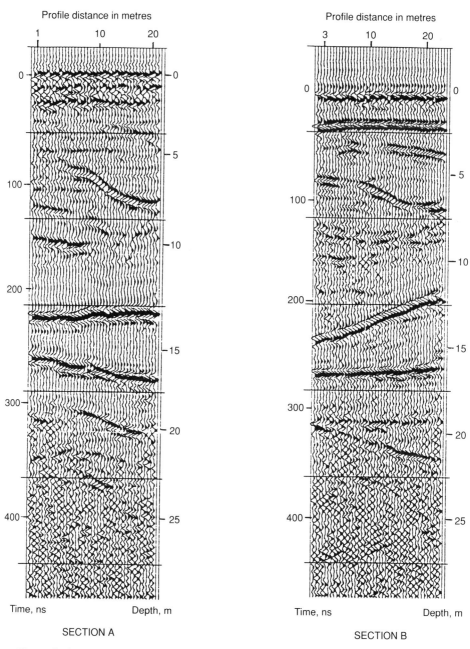

Figure 7.17 Radar measurement of granite (courtesy of CEA France)

out in the laboratory on rock samples cut in different directions in relation to the schistosity. These schists are an extreme case of anisotropy but the tests that were made prove that particular attention should be given to velocity measurement in a complex site.

Besides the effects of anisotropy on wave propagation, the 'volumic scattering' effect due to the heterogeneity of the massif was determined. To do so, data were collected according to different geometrical configurations of the antennas (antennas parallel or perpendicular to the measurement line, transmitter parallel or perpendicular to the receiver). The terrain measurements were completed by data processing using a program perfected for seismic studies concerning shear-wave birefringency. The complexity of prospecting on a site where volumic scattering occurs was shown by comparing the radar response recorded when the antennas set perpendicular to each other with that recorded when the antennas were arranged colinearly: with no volumic scattering, no reflection is collected with perpendicular antennas whereas when volumic scattering occurs, the heterogeneities reflected the waves differently according to their position in relation to the antenna planes E and H.

These tests give a general idea of the studies scheduled to determine the investigation possibilities offered by radar techniques in geophysics: the diversity of the studies that must be undertaken in order to draw as much information as possible from a radar measurement campaign is large. The lessons drawn from our preliminary tests made on several very different sites with surface equipment should be of use for research on the technique implemented in boreholes or in galleries in configurations representative of those encountered in waste-storage-site qualification work. Further effort has to be invested in signal processing so as to achieve dynamic hydraulic characterisation of a granitic formation.

7.4.6 Snow, permafrost and glaciers

An early use of radar to probe into glaciers was by Steenson (1951). Steenson concluded that radar techniques were a viable method of measuring ice thickness. Cook (1960) proposed an airborne system and many surveys have been subsequently carried out from helicopters and fixed wing aircraft.

Waite and Schmidt (1962) found that pulsed radar altimeters gave erroneously low readings over ice and snow in the Antarctic and suggested that some accidents may have been caused by incorrect readings.

In the UK in 1963, at the Scott Polar Institute, Cambridge, Evans (1965) used radar techniques to measure ice thickness and glacier depth in the Antarctic. There followed much work by researchers from the USSR, USA and UK in the Antarctic investigating the ice and snow cover. A popular review of

Antarctic ice work is by Radok (1985) and there is another by Kovacs and Morey (1989) in one of a series of reports for the US Army.

Biggs (1977) reported on the profiling of the Great Lakes using a 2.8 GHz short-pulse radar which was initially mounted on an aircraft and subsequently towed above the ice surface.

Hall (1982) described the use of a short-pulse radar of centre frequency 480 MHz, mounted on a helicopter, to perform an aerial survey of the route of the Alaskan pipeline. The radar profiled the permafrost to depths of 3 m. Permafrost was also investigated by Unterberger (1978) who used an airborne radar of conventional design. This operated at a centre frequency of 250 MHz with a pulse length of 2 μs, radiated by a ten-element Yagi antenna array. Working in Alaska at Umiat, he probed up to 600 m in permafrost test wells. Rodeick (1983) showed the clear detection of a 'thaw bulb' in permafrost using a 120 MHz antenna and an impulse radar.

Work in the USSR involved radar probing of ice in the Siberian sea and glaciers on the island of Bendetta. Using an airborne radar operating at 38 MHz, and with phaseshift-keying modulation, Lukashenko *et al.* (1979) (USSR) probed through ice formations. From a height of 1500 m the radar detected reflections at ice depths of up to 120 m.

Finkelstein (1977) (USSR) reviewed a number of applications of airborne radar to investigate sea ice and freshwater ice. By means of a system operating in the frequency region 20–130 MHz, measurement of sea ice thickness up to 2.5 m was achieved. Annan and Davis (1977) also describe measurement of ice thickness and freshwater bathymetry. Recent investigations of seawater ice have been made by Cambridge Consultants Ltd (UK), and aspects of this work are reported by Oswald (1988).

Work by Bishop *et al.* (1980) resulted in probing the Vatnajökull icecap in Iceland to a depth of 600 m using 2 MHz radiation transmitted from 60 m resistively loaded wire antennas. Dong *et al.* (1980) reported probing of the Hispar glacier.

Stratigraphy of snow has been the subject of a number of studies, including those by Ellerbruch and Boyne (1986) and by Fujina *et al.* (1985). These involved FMCW systems, Fujino *et al.* using the frequency range 2–12 GHz. Such frequencies, high by most subsurface-radar standards, are possible in snow probing because of the low loss of frozen water, and because its density is low (mean relative permittivity in the region of 1.5).

An Anglo–Swedish team investigated glaciers using a pulsed radar with a central frequency of approximately 60 MHz. This work was carried out in the Swedish Arctic at Storglaciaren. Successful probing of the glacier at depths of up to 40 m was achieved. Deeper penetration was obtained with a radar working at 5 MHz. Multifrequency (8–10 GHz) holographic investigation of snow has recently been reported by Sakamoto and Aoki (1985) with a view to location of objects buried by snow avalanches. Similar investigations have been carried out by Daniels as a means for detecting skiers engulfed in avalanches. A picture of one of the volunteers is shown in Figure 7.18.

Figure 7.18   Detecting skiers trapped in an avalanche (trials) (courtesy of ERA)

7.5 Archaeological applications

7.5.1 English Heritage, UK
J. Fidler, RIBA

Surface-penetrating radar has been used to locate voids, inconsistencies and buried metalwork in a wide variety of structures: mediaeval cathedrals, castles, Egyptian pyramids and even the Ark. However, certain requirements need to be considered.

Geophysicists and electrical/radar engineers are not architects, structural engineers or building surveyors and their knowledge and understanding of construction systems and faults is understandably limited, which affects their interpretation of the radar images. The building professionals' hopes and expectancies for remote sensing outstrip their perception of the actual capabilities of radar techniques. Only by having both sets of professionals working as a team on site and subsequently in the office can misinterpretation be eliminated. It is not acceptable for one group to survey and interpret the building and for the other to receive and accept the results.

Radar is best employed alongside magnetometers, resistance and radio-detection devices etc., with a thorough-going visual assessment of the building. The best surveys establish an hypothesis of construction and deformation/decay. They model this hypothesis off the building, by creating freestanding facsimiles in cross-section which can be compared with the radar signals, and vary configurations identified to unique signatures so that data from the actual building can be properly interpreted.

Typical radar applications include identifying the scope and extent of voids in medieval cored walls where dressed stone ashlar faces are supposed to be tied and consolidated by a lime/rubble fill. In ruined abbeys and castles, where walls are saturated, the cementing lime concrete dissolves and leaches away, freeing rubble masonry to move under gravity, creating instability, bulging and cracking of the structure.

7.5.2 Wharram Percy, a medieval village (ERA Technology, UK)

D. J. Daniels

An example of the use of radar in archaeological exploration is in Yorkshire (UK). Wharram Percy is an example of an abandoned feudal medieval village

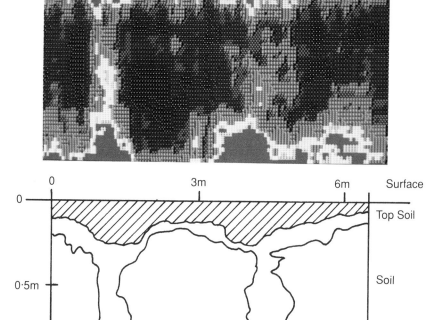

Foundations

Figure 7.19   Radar survey of a medieval building at Wharram Percy, UK (courtesy of ERA Technology)

near York. In conjunction with the University of York, radar surveys were carried out at Wharram Percy.

Wharram Percy may have started as a group of scattered farms which gradually coalesced to form a village served by a Church. This was probably established c.800 AD and is now known as St Martin's Church and dates from 1200 AD. The village houses were generally well built and were constructed from chalk rubble, wattle and daub.

Wharram Percy is located about a small river valley on a north–south axis with a pond for fish stocks, a water mill and in mediaeval times a line of buildings in the valley for occupation by the peasants. On the west ridge above the valley the Manor House was situated and adjacent to this building were various halls and smaller buildings with chalk foundations.

A radar survey over one of these buildings is shown in Figure 7.19 and the buried foundations can be deduced from the radar image.

7.5.3 Fountains Abbey (ERA Technology, UK)

D. J. Daniels

Fountains Abbey, in North Yorkshire, founded in 1134 by the Cistercians, is an outstanding example of a medieval monastery. It stands in the valley of the River Skell and the existing ruins give a clear impression of life in

Figure 7.20   Fountains Abbey

Figure 7.21   The Cellarium, Fountains Abbey

one of England's greatest and best preserved abbeys. Although the first construction was of timber, the Abbey grew to encompass a stone built church 360 ft long with a 168 ft tower. The Cellarium with its double row of arches is an inspiring reminder of monastic life. Photographs of the Abbey are shown in Figures 7.20 and 7.21 and a plan is shown in Figure 7.22.

Recently, and in collaboration with the University of York's Department of Electronics, ERA Technology carried out a survey of parts of the grounds of the Abbey. Of considerable interest is the site of a Guest Hall, no longer in existence, and examples of a radar survey are shown in Figure 7.23. In this figure the radar image shows two main features. The strongest feature on the righthand side, and also shown in Figure 7.23*b* is the remains of a pillar which is conjectured to be a support for part of the Guest Hall. Also to be seen are horizontal lines which are believed to denote the position of the original floor of the Guest Hall. The key features are the remains of pillars and the contours of earlier floor levels.

Other features which were detected under the Cellarium and the Infirmary Hall were the culverts directing the River Skell, which in previous times functioned as the water system. These are shown in Figure 7.24. The radar image of the culverts under the Cellarium is shown in Figure 7.24*a* and an interpretation is shown in Figure 7.24*b* which is a typical example of how much detail a radar image can provide of underground features. The radar image in this particular case is unfocused and therefore most of the reflections give rise to

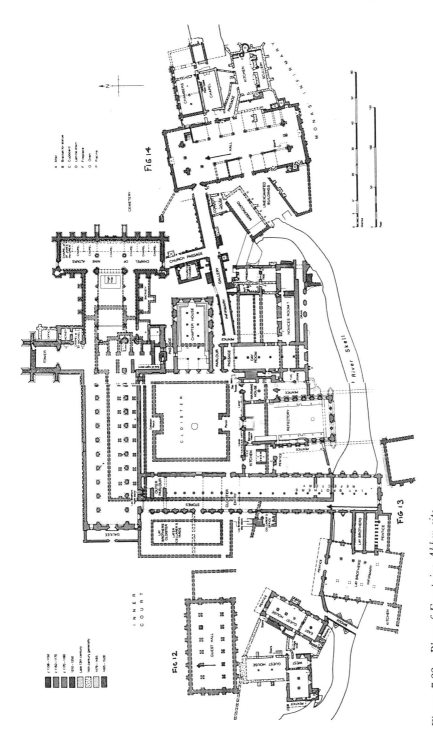

Figure 7.22 Plan of Fountains Abbey site
Source: Coppack *et al.* English Heritage 1993

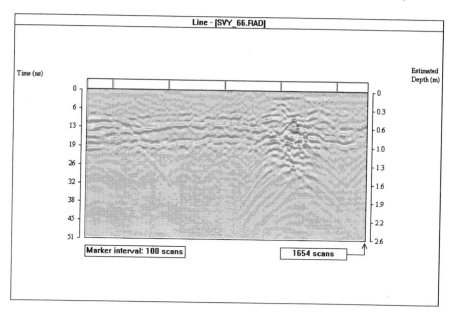

Figure 7.23a   Radar image of the Guest Hall (courtesy ERA Technology)

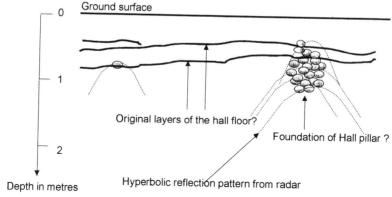

Figure 7.23b   Interpretation of radar image (courtesy ERA Technology)

hyperbolic traces. This is particularly noticeable when caused by the corner reflectors formed by the river water and the masonry side wall of the culvert. Other hyperbolic reflectors can also be seen and these are caused by the masonry side supports of the culvert. (ERA Technology gratefully acknowledges the assistance of English Heritage in permitting access to the archaeological site at Fountains Abbey which is owned by The National Trust and in the care of English Heritage.)

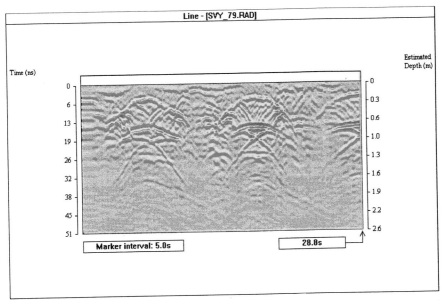

Figure 7.24a *Radar image of the culverts under the Cellarium (courtesy ERA Technology)*

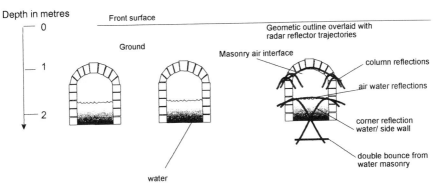

Figure 7.24b *Interpretation of radar image (courtesy ERA Technology)*

7.6 Civil engineering application

7.6.1 Tunnels (ERA Technology Ltd, UK)

D. J. Daniels

During the 19th century, when the railway system was the dominant means of mass transport, many tunnels and bridges were built as the rail network spread throughout Europe. These structures were generally made from brick and owing to the high level of safety margin generally used by railway constructors, have survived well for over more than a century. The railway network in the UK still regularly passes through brick lined tunnels and over brick bridges. In the UK the outstanding pioneers of railway expansion were the Stephensons, both father and son, and Isambard Brunel and their engineering and management skills still inspire admiration.

Some idea of the scale of the undertakings can be gained from the statistics of the work involved in building the railway line from London to Bristol in the early part of the last century. In 1836 Brunel started work on the longest tunnel then ever attempted, of some two miles (3.2 km) at Box. Six permanent and two temporary shafts of 8.5 m diameter were sunk from hilltop to railway level. The deepest shaft was nearly 93 m deep and when the shafts were complete, work on the tunnel excavation proper was started by the contractors. This excavation ran through mixed conditions: clay, blue marl, inferior oolite and oolite, and was eventually lined for 75% of its length with bricks. Gunpowder was used to blast away the rock which formed 25% of the tunnel but the remaining length was excavated by manpower and horsepower working in candlelight. Over the two and a half years taken to complete the tunnel excavation, a ton each of gunpowder and candles was used each week. Some 30 000 000 bricks were used to line the tunnel and over a quarter of a million cubic yards of soil was hauled out of the shafts in buckets by horsepower. The tunnel was completed in June 1841 although, even in those days, the final cost of tunnel construction was over twice the original estimates.

The safety record of such structures has been excellent, not only because of their intrinsic strength but also because of regular inspection procedures which have been carried out for many years.

In the UK, the railway operators may wish to gather information on the integrity of the tunnel linings by inspecting both their internal condition and the depth of the masonry lining, and has therefore identified a need for an economical, rapid, nondestructive technique for tunnel-line inspection.

Several examples of tunnel inspection using radar follow the first example of a bridge abutment.

A photograph of the abutment of the bridge is shown in Figure 7.25a. The total area of the abutment wall examined was 3×6 m. A grid of 100 mm spacing was used in collecting data and this produced an array of 30×60 time-domain waveforms for processing.

Each x, z plane of data was suitably processed and it can be seen from the processed image in Figure 7.25b that there is a strong radar reflection in the centre of the scan (horizontal axis 6 m, vertical axis 1.5 m). This image (by virtue of its low-frequency energy content) was successfully interpreted to indicate an anomalous water saturated zone. Subsequent trial core samples confirmed the interpretation.

To generate radar images from known situations, a test rig with suitably simulated voids and anomalies 12.5 m long, approximately 1 m deep and 1.5 m high was built, as illustrated in Figure 7.26. The test rig comprised three

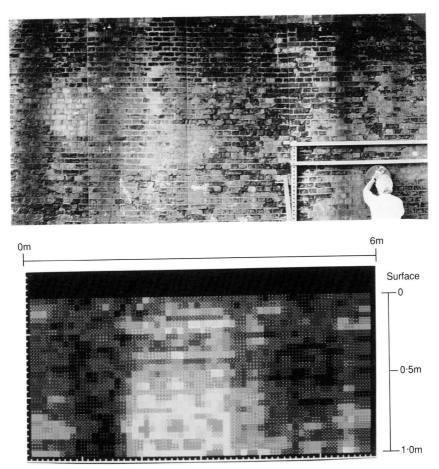

Figure 7.25   Radar survey of bridge abutment (courtesy of British Rail)

continuous sections: the first section was designed to simulate various thicknesses of tunnel lining (varying from one brick to five bricks), the second to simulate vertical and horizontal delaminations, and the third to simulate various voids.

The test rig was subsequently modified by extending the depth using a sand backwall. The void is in the centre of the radar image in Figure 7.26.

Radar measurements were taken on Gonerby Tunnel which lies approximately 1250 m north-west of Grantham (UK) (Long 0° 41′W and Lat 52° 55.5′). The tunnel runs west north west, is approximately 520 m long and serves a dual track railway line. The tunnel was constructed in Victorian times and passes through lower lias clay with calcareous siltstones and thin sandstones and underneath middle lias, grey, sandy clay and micaceous clay 15–30 m thick. The roof of the tunnel lining varies in height from the railway line by typically ±0.75 m and in addition there is a step reduction in height at a distance of some 80 m from the south entrance of the tunnel. Photographs of the tunnel and its

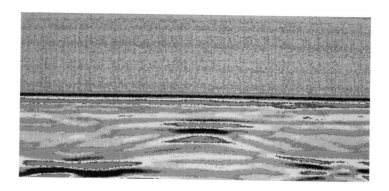

Figure 7.26   Radar survey of tunnel-lining test rig (courtesy of British Rail)

Figure 7.27   Railway tunnel (courtesy of British Rail)

Figure 7.28   Tunnel lining (courtesy of British Rail)

14m

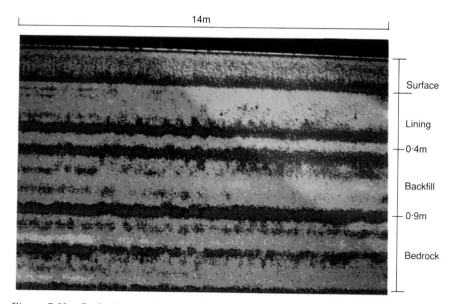

Surface

Lining

0·4m

Backfill

0·9m

Bedrock

Figure 7.29   Radar image of clear section of tunnel lining (courtesy of British Rail)

lining are shown in Figures 7.27 and 7.28. The tunnel was surveyed at the crown of the roof and a continuous radar image was recorded. A typical radar image of an undisturbed region is shown in Figure 7.29 and the thickness of the lining was calculated to be 0.56 m. A typical image of a disturbed region is shown in Figure

Surface

Lining

0·4m

Backfill

0·9m

Bedrock

Figure 7.30   Radar image of disturbed section of tunnel lining (courtesy of British Rail)

7.30. From this information, a composite map of regions of high-level radar reflection was produced.

7.7 Roads

7.7.1 Introduction (Durham County Council, UK)

D. L. Wilkinson

Roads or pavements can be constructed in a number of ways and can be classified into flexible or rigid types. Flexible construction uses bituminous materials and rigid construction includes a concrete layer, which may be unreinforced or reinforced.

Virtually all road constructions are built up with layers of different materials.

One of the major indicators of condition is the strength of the road pavement, and this is measured by the deflectograph. This is a heavy vehicle which measures the deflection of the pavement under the transient loading of its rear wheels. These deflections are then analysed, together with other data such as traffic flow and pavement thickness and condition, to provide an estimate of the residual life of the pavement. The deflections can also be used to calculate an overlay thickness to provide a pavement with a given future life.

Radar techniques may be used to determine the thickness of bituminous layers and concrete slabs, the spacing and location of slab reinforcement, their depth of cover, dowel bars joining slabs, cracking and for unreinforced or widely spaced reinforcing, the position of cracks beneath slabs (Figure 7.31).

Pavement thickness has traditionally been determined by drilling and extracting cores, and pavement condition by excavating trial holes. Disadvantages of these methods are damage to the road structure and provision of information only at widely spaced points. In the early 1980s it was realised that ground-penetrating radar might be able to provide this information and a few pavement engineers tried the technique with some success. In addition to establishing pavement-layer thicknesses, they were finding that the technique was useful in establishing ground conditions below the pavement layers; a task that can be performed by trial pitting but is expensive.

By 1990 a number of radar surveys had been carried out in the UK and the technique promised to be a success. However, the experience was fragmented and the requirements of radar surveys varied from scheme to scheme and between pavement engineers. Radar providers needed a better perception of the pavement engineer's requirements for the technique so that developments in equipment could be concentrated on what was needed. After consultation, the County Surveyor's Society, through its Data Collection Working Party, produced 'Guidelines for the development of ground radar' in 1991. The guidelines included requirements for accuracy, calibration, robustness, the

Figure 7.31   Radar crack-detection system (courtesy of TRRL)

ability to locate results to specific positions on the road, and computerisation of results.

Secondary requirements were seen as ability to differentiate between sound and broken materials in the bituminous layers, and the presence of defects such as voids, delamination between layers, cracking of the surface layer and moisture content variation in the granular layers. Since the guidelines were published the technique has moved on. Accuracy has increased to the point where it matches coring. Repeatability and reproducibility are acceptable. Operators are offering high-speed surveys with results referenced to the road network and low-speed surveys where more detail is required.

Demonstrations of equipment have been given to the Data Collection Working Party and the Transport Research Laboratory. The DCWP demonstration has shown that accuracy and agreement on interface depths

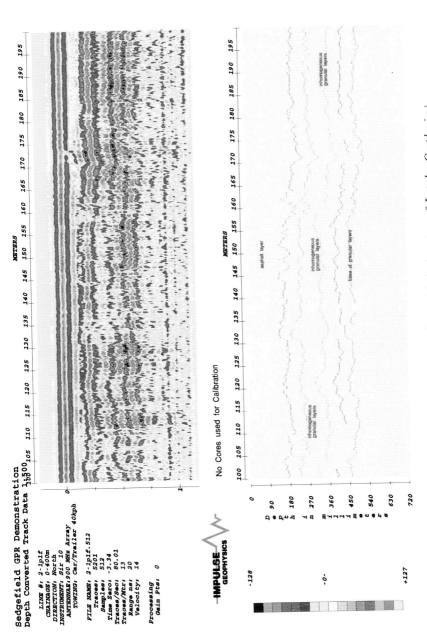

Figure 7.32 Test results from UK trials of radar for road surface analysis (courtesy of Impulse Geophysics)

between different operators is good on roads with a predictable construction. However, apart from motorways, trunk roads and major 'A' class roads, most roads have been built up over the years in a piecemeal fashion. It was found that on this type of road operators agreed on the major interface depths, but were sometimes unable to agree on material types and picked up varying minor interfaces (e.g. thin layers of regulating material, internal layers within, for example, a subbase). On this type of road it would, therefore, be essential to correlate radar surveys with road coring and, possibly, trial pitting. It has to be said that roads with a predictable construction, although comprising only about 20% of the total road network, receive a large proportion of maintenance funding—perhaps 80%.

The DCWP demonstration was mainly of low-speed equipment but one operator demonstrated his equipment at high speed. He obtained results of comparable accuracy with the best of the low-speed-equipment operators. This demonstration, and trials carried out elsewhere, have shown that high-speed surveys offer an opportunity to survey long stretches of the major road network without the need for traffic control. This is an important consideration because traffic control is expensive and causes delay and danger to motorists. Various test results are shown in Figures 7.32 and 7.33.

The improvement in accuracy of ground-penetrating radar over the past few years has been encouraging and increased use of the technique should lead to further improvements. A survey in 1994 found that 32 County Councils had used ground penetrating radar to solve a variety of pavement and structural problems.

Further use of ground radar by local authorities is being held up by the lack of a specification for the work and by the difficulty authorities unused to the technique have in selecting operators for a tender list. In 1994 the Department of Transport issued an Advice Note (HA 72/94) entitled 'Use and Limitation of Ground Penetrating Radar for Pavement Assessment'. This document is of assistance when preparing a ground penetration radar survey contract.

7.7.2 Laboratoires des Ponts et Chausées, France
Drs. J. Cariou, Ph. Côte and X. Derobert

In France, considerable sums of money are spent in maintaining existing roads to enable them to carry the ever increasing traffic flows and loadings. Funding is largely provided by central government and the highway maintenance engineer must spend the money in the most cost effective manner by targeting those roads most in need of treatment. To do this he must know the condition of the existing road network. References to methods of non destructive examination techniques can be found in Ransome (1986), Holt (1987), Clemena (1987), Bomar (1985), Kunz (1988).

During the last ten years, the Laboratoires des Ponts et Chaussées (LPC) has

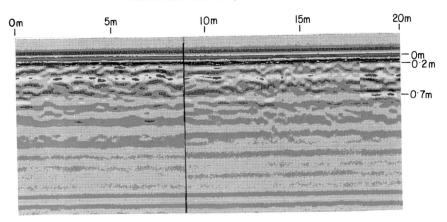

Processed data - Road Survey Section 0 - 20m

Processed data - Road Survey Section 110 - 130m

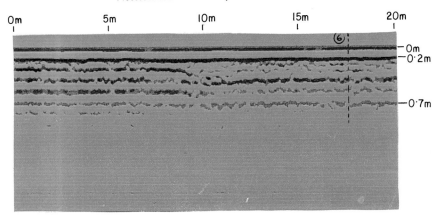

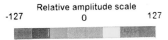

Relative amplitude scale

-127   0   127

Figure 7.33   *Test results from UK trials of road-survey radar (courtesy of ERA Technology)*
   a Processed data, road survey section 0–20 m
   b Processed data, road survey section 110–130 m

been investigating two applications of GPR: first for road inspection and secondly for geological and geotechnical surveys.

The main application of the LPC road-inspection radar is the measurement of thicknesses of pavement structures (bituminous layers and treated base). LPC

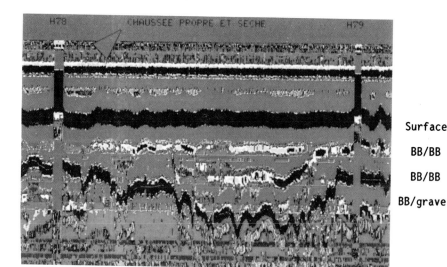

Figure 7.34   *Radar profile of road (courtesy of LPC)*

has developed software to calibrate the equipment and to improve the signal-to-noise ratio.

The radar is a monostatic type system with a wide-frequency antenna centred around 700 MHz. It gives simultaneously information on the thickness of the first two layers and a pseudo-colour line-scan representation. The processing time is 20 ms, so at the maximum speed 20 km/h a spatial survey every 10 cm can be achieved. A typical test result is shown in Figure 7.34.

The minimum thickness that can be measured is about 5 cm with an accuracy within approximately 8%. This accuracy is confirmed when using corings to determine the speed of the electromagnetic wave in the material. Qualifying the bonding between courses is a more difficult problem, particularly between bituminous layer and base.

During the late 1980s, LPC experimented with synthetic-pulse-radar systems in collaboration with two other French companies (SNCF and RATP). The choice of this new type of radar derived from a requirement to improve the performance of existing methods for which the depth of investigation is sometimes insufficient.

The results obtained confirmed the potential of synthetic-pulse radars and demonstrated that they can exceed the dynamic range obtained with current time-domain radars. One consequence of this type of modification could be an increase of the investigation depth by a factor of 1.5 and possibly more. LPC is now developing a synthetic-pulse radar system.

In 1992 a GSSI system was used for general surveys and experiments on special problems. LPC is at present (1995) developing research in water-content-measurement of road layers (treated bases) and for debonding detection.

To detect underground cavities, LPC has designed a method based on

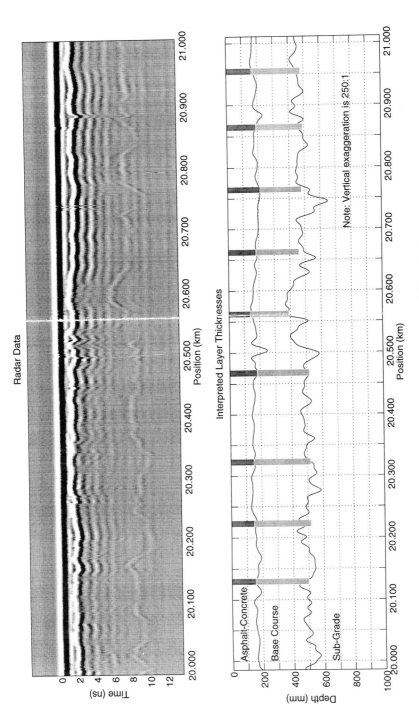

Figure 7.35   Radar analysis of asphalt and base-course thicknesses (courtesy of Road Data Ltd)
Note:Radar interpretation was completed without the benefit of core information; cores were taken after the fact to confirm results

interaction between a monochromatic EM wave, in the frequency range 100 MHz–1 GHz, and the ground situated between two boreholes. The equipment permits tomographical interpretations for distances between boreholes up to 10 m with a frequency of 200 MHz.

In the geotechnical field, LPC is developing new treatments, with the GSSI device, derived from seismic techniques, in order to improve performances such as common mid-point, common depth point and 3D interpretations.

General

Many methods of road survey rely on comparison with coring. However, a multichannel radar system can be used to provide calibration and hence immediate measurements of road-layer thickness as shown in Figure 7.35. This system was developed by Road Radar (Canada) for rapid and accurate road and bridge survey.

7.8 Remote sensing

Subsurface imaging by radar from satellites is possible where the topographic cover is radar smooth and the material penetrated is fine grained, no more than a few metres thick and very dry. Where the thickness of the cover is less than the skin depth, the return signal can be significantly increased because of refraction of the electromagnetic wave, and reduction of backscatter due to oblique incidence.

Kadaba (1976) has reported data on the penetration of 0.1–1.5 GHz radiation into the earth's surface for remote-sensing applications. Elachi *et al.* (1984) (Jet Propulsion Laboratory, USA) demonstrated that the satellite radar SIR-A (shuttle-imaging radar) could detect dry river beds beneath the Sahara Desert. This ability is due to the hyperarid conditions existing in desert regions where the skin depth of the sand can be 5 m or more. Blom *et al.* (1984) (USA) describe the detection of igneous dykes beneath the Mojave Desert. The *Seasat* satellite L-band radar provided the basic data to generate the image. The dykes are buried under 1–2 m of alluvial cover. Recently, a range of other reports has appeared of SIR-B results of features covered by desert sand.

The *Apollo 17* Lunar Sounder experiment was described in detail by Brown (1971) before the launch. Both amplitude and phase characteristics of the echo were to be measured so as to extract maximum information from the measurements. Surface-clutter-reduction techniques were to be used.

Porcello *et al.* (1974) detailed the results of the mission, and showed that the *Apollo 17* experiment could provide profile information on subsurface geologic structures. This is possible on the moon because of the very low attenuation of electromagnetic radiation found in lunar rocks. Using a synthetic-aperture radar operating at 5, 15 and 150 MHz they photographed the radar data in conventional SAR form and returned them to earth for processing. The 150 MHz VHF system was used for imaging the Mare Crisium. The same area was investigated by Maxwell and Phillips (1978) (USA) who detected interfaces at depths of 1400 m and 1000 m in the central region. These results were achieved using a 5 MHz radar.

7.8.1 Earth (United States Department of Interior, USA)

Dr. G. Schaber

The first NASA Shuttle Imaging Radar (SIR-A and SIR-B) images acquired over Egypt and Sudan demonstrated the capability of spaceborne 24 cm-wavelength (L-band), synthetic-aperture radar (SAR) signals to return geological information from depths of 1–2 m within the loose sand cover and the semiconsolidated alluvium typical of this desert surface (Schaber *et al.*, 1986, 1990); see Figure 7.36.

Images produced by the return of SIR signals from beneath the surficial sediment reveal patterns of previously unknown, fully aggraded, stream valleys and erosional surfaces that are at present mantled by a thin sand cover (McCauley *et al.*, 1982; Breed *et al.*, 1983).

The importance of this discovery lies in (i) its application to new interpretations of the Cenozoic geology of the northeastern Sahara, and (ii) recognition of the possible use of these fluvial pathways during the late Quaternary by Stone Age people. These questions have been explored by geologists from the US Geological Survey in collaboration with geologists, geomorphologists and archaeologists from the USA and Egypt (Breed *et al.*, 1982; 1987; McCauley *et al.*, 1986a; 1986b; McHugh *et al.*, 1988a; 1988b; 1989; Szabo *et al.*, 1989; Issawi and McCauley, 1992).

The desert in southwestern Egypt and northeastern Sudan, about the size of the US state of Arizona, is uninhabited and was poorly known until about 20 years ago. It was considered unusual because it lacks surface drainage features; instead it has an almost ubiquitous and complex Quaternary eolian veneer. The scattered, extensively wind eroded, decameter-scale outcrops do not display clear fluvial patterns either on the ground, or on conventional visible or near infrared wavelength images such as Landsat and SPOT. The radar sensor's portrayal of the buried river valleys and their surrounding relict fluvial topography came about because the surface sand in this part of the Sahara is underlain by both a regional duricrust of secondary calcium carbonate (caliche or calcrete) deposited in the sediments of alluvial valleys during periods wetter than the present (Schaber *et al*, 1986); and by bedrock terraces that mark the shores of the old alluvial valleys (McCauley *et al.*, 1986a). The deflated, hard caliche substrate now remains as a regional carapace that mimics and preserves the patterns of successive geomorphic surfaces and associated streamcourses that episodically occupied the buried valleys between early Tertiary and middle Holocene time (McCauley *et al.*, 1986a). The probing Shuttle Imaging Radar sensor revealed these features through the thin, sand-sheet cover.

Subsurface imaging of natural terrains using long-wavelength SAR from spaceborne (and air-borne) platforms has been demonstrated only where the

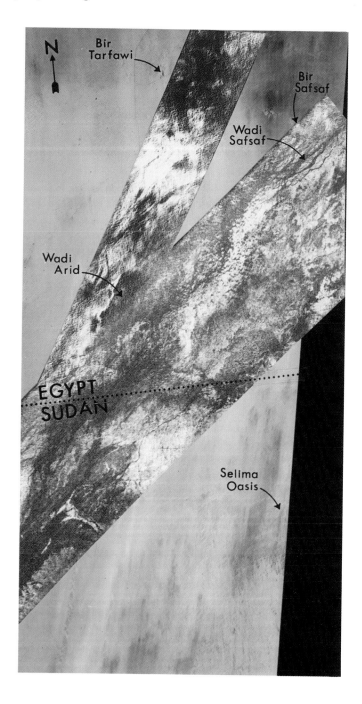

Figure 7.36   SIR image of desert landscape (courtesy US Dept. of Interior)

topographic surface is radar smooth (dark on the radar image), and where the shallow subsurface contains a dielectric interface or disseminated dielectric inhomogeneities that are 'radar-rough' (bright on the radar image). For successful radar imaging of the shallow subsurface, the penetrated material must be fine grained, no more than a few metres thick and extremely dry. The single most important factor in the SIR portrayal of subsurface geology in the northeastern Sahara is a change in loss tangent (tan δ, which reflects a change in the imaginary part of the permittivity) due to scattering in dry and clay-free alluvial materials underlying a thin and equally dry sand sheet (Schaber *et al.*, 1986).

Similar radar penetration from space altitudes has been demonstrated in the Badan-Jaran Desert of China (Breed *et al.*, 1982; Guo *et al.*, 1986), in Saudi Arabia, Berlin *et al.* (1986) and in the Mojave Desert of California (Blom *et al.*, 1984; Farr *et al.*, 1986). Long-wavelength SAR subsurface mapping using multifrequency airborne SAR sensors had been demonstrated in the deserts of California (Ellermeier *et al.*, 1967; Dellwig, 1969), and later in the deserts of southwestern and north-central Arizona (Schaber and Breed, 1993).

Schaber *et al.* (1986) found from field studies in southern Egypt (including the study of 150 hand-dug pits and 82 back-hoe trenches) that the maximum 'radar-imaging depth' (RID) for 24 cm-wavelength (L-band) SIR signals to a buried dielectric interface in the sandy alluvium typical of the hyperarid Western Desert of Egypt could be expressed as

$$RID = 0.25(1/\alpha)$$

where $1/\alpha$ is the electrical skin depth, or attenuation length. The field attenuation coefficient (α) for low-loss media ($1 \gg \tan\delta$) is $\alpha = \frac{1}{2}k\tan\delta$, and k is the wavenumber (Ulaby *et al.*, 1982).

Schaber *et al.* (1986) further found that dense, dry caliche nodules disseminated in the upper part of the sandy alluvium in the Western Desert of Egypt result in strong volume scattering. Such scattering enables discrimination on the SAR images between the radar-bright regions between small fluvial channels and the channels that are radar-dark due to mirror-like reflection from the flat, gravel-rich surfaces below a very thin sand mantle.

Using SIR-A results in the Sahara reported by McCauley *et al.* (1982), Elachi *et al.* (1984) showed, theoretically, that the presence of a thin low-loss (dry) sand layer as thick as two-thirds of the attenuation length (or skin depth) will in effect enhance the radar's capability to image the subsurface interface with HH polarisation. A similar sand layer as thick as one-quarter of the attenuation length will enhance the radar's capability to image the subsurface interface with VV polarisation. For an attenuation length of 6 m, Elachi *et al.* (1984) found that this favourable effect will occur for sand layers up to 4 m thick for HH and up to 1.5 m thick for VV. They also showed that, even though the absolute backscatter cross-section decreases as a function of incidence angle (θ_i), the presence of a dry sand layer and the resulting refraction effect can enhance the capability to image the subsurface, particularly at large incidence angles (about 50° or greater).

The radar images obtained of the Eastern Sahara during SIR-A and SIR-B, though few, indicate that this remote sensor can be applied to regional exploration for shallow ground water associated with sand-buried paleovalley systems (McCauley *et al.*, 1986*a*; 1986*b*). Radar can also be used to map the distribution of certain types of duricrusts, such as calcrete, ferricrete and gypcrete, which are diagnostic of various paleoclimatic conditions. The radar data can be used synergistically with information from other types of remote sensing (such as spectral data from Landsat or SPOT multispectral images) to provide simultaneous information on both the surface and subsurface, of desert regions (Davis *et al.*, 1993).

As part of NASA's third Shuttle Imaging Radar Mission (SIR-C/X-SAR) (two launches in 1994), acquisition of additional SAR image data over the Sahara, the Arabian Peninsula and other major deserts, was planned; this will help reconstruct the scope and trends of major paleodrainage systems. Another objective will be to evaluate further the optimal sensor and geologic parameters conducive to subsurface mapping and delineation of duricrusts in desert terrains, which are expected to provide evidence for regional changes in paleoclimate. The SIR-C/X-SAR platform includes an X-band (3 cm-wavelength) SAR (VV polarisation) developed by the German Space Agency and the Italian Space Agency and fully polarimetric (HH-HV, VV-VH) C- (6 cm) and L-band (23 cm) SARs developed by NASA's Jet Propulsion Laboratory (Pasadena, California).

7.8.2 Extra-terrestrial

7.8.2.1 Mars 96 Mission (Max Plank Institut für Aeronomie, Germany) Prof. T. Hagfors

The Max Plank-Institut für Aeronomie (Professor Hagfors and Dr Nielsen) are contributing to the Russian Mars 96 mission which is due to start in the late autumn of 1996 and is intended to make extensive measurements of particles, composition and fields in the atmosphere of Mars and to make maps of the surface of Mars from a satellite in orbit round the planet. The instrumentation includes a long-wavelength radar designed as a topside ionosonde to explore the upper ionosphere of the planet, a region which has previously only been investigated in occultation experiments and in brief periods during the descent of Mars probes. This topside sounder is a swept-frequency radar which operates between 170 kHz and 5 MHz.

During night time on Mars the maximum ionospheric plasma frequency falls well below 1 MHz, and a wide frequency window opens in which the observed echoes are from the martian surface rather than from the ionospheric plasma.

The use of this window to explore the surface offers exciting new possibilities for the study of Mars. Except for the polar caps, the surface has the appearance of a dry desert. The observation of ancient riverbeds and areas which must have been flooded, however, suggests that the surface of Mars cannot always have been so. Indeed, water must have flowed in abundance at some time in the past to create the fluvial features. This water cannot all have escaped from the surface, and current opinion holds that much of the water has been captured and is held under the surface in the form of ice, and that the depth of this ice is a function of martian latitude. There is also observational evidence, based on the shape of some of the surface features of Mars, that ice must exist under the surface. Estimates of the depth of these ice, or icy layers range from a few tens to a few hundreds of metres.

This layered structure of the surface must lead to quite characteristic variation of the reflection coefficient with frequency. The simplest model of such a frequency variation is provided by a dielectric medium of thickness L and relative permittivity ϵ_1 resting on top of a semi-infinite region with relative permittivity ϵ_2. In this case the power reflection coefficient of the surface at normal incidence takes the form

$$R^2 = \frac{R_1^2 + R_2^2\, e^{-2\tau} + 2R_1 R_2\, e^{-\tau} \cos\theta}{1 + R_1^2 R_2^2\, e^{-2\tau} + 2R_1 R_2\, e^{-\tau} \cos\theta}$$

with

$$\theta = \frac{4\pi f}{c}\left(\sqrt{\epsilon_1'}\right)L$$

$$\tau = \frac{2\pi f}{c}\left(\sqrt{\epsilon_1'}\right)L \tan\Delta_1 = \frac{\theta}{2}\tan\Delta_1$$

where θ is the phase path through the upper layer and back and where Δ_1 is the ratio of the imaginary and real parts of the relative permittivity. The reflection coefficients are defined as for semi-infinite half space as

$$R_1 = \frac{\left(\sqrt{\epsilon_1'}\right) - 1}{\left(\sqrt{\epsilon_1'}\right) + 1}$$

$$R_2 = \frac{\sqrt{\left(\epsilon_2'/\epsilon_1'\right)} - 1}{\sqrt{\left(\epsilon_2'/\epsilon_1'\right)} + 1}$$

In the absence of absorption in the upper dielectric, the reflection coefficient will oscillate with frequency between the two values

Table 7.2   Mars 96 LWR experiment parameters

Frequency range	0.17–4.93	MHz
Transmitter bandwidth in		
frequency band 1: 0.18–2.1 MHz	2660	Hz
frequency band 2: 1.9–5.0 MHz	665	Hz
Number of discrete frequencies	56	
Range of linear frequency modulation	15	kHz
Pulse repetition frequency	300–900	Hz
Time of measurements at each frequency	270 or 1080	s
Dipole antenna length	2×20	m
Peak voltage at antenna input	1.2	kV
Power	100	W
Total mass	35	kg

$$\text{MaxMin} = \frac{(R_1 \pm R_2)^2}{(1 \pm R_1 R_2)^2}$$

with the thickness determining the period of the oscillation in the frequency domain.

It is, of course, unlikely that the real situation will be as simple as this model. There may be a gradual transition in the relative permittivity with depth, the thickness of the layer may change in a random fashion with position on the surface and there may be substantial attenuation in the surface layer. The purely oscillating behaviour of the reflectivity will then probably be replaced by a gradual and nonoscillating behaviour of the reflection coefficient. Even so, the form of the transition will identify the thickness of the layer and the dielectric properties of the surface material, and may serve to confirm the presence of a buried ice layer on Mars.

The properties of the orbit and the radar system carried in the Mars 96 satellite are listed in Table 7.2.

The radar system will feed into a dipole antenna consisting of two 20 m retractable booms. The radar equipment is integrated into the spacecraft which is shown in Figure 7.37. The spacecraft was scheduled to be launched in Autumn 1996 and to be injected into orbit the next year.

7.8.2.2 Mars 96 Project (University of Rennes, France)
F. Nicollin[*1]

One of the instruments of the Mars 96 balloon mission is a Ground Penetrating Radar system, which is planned to be integrated inside the guiderope (ballast of

* This work was supported by CNES (MARS 96 balloon project and Radar Research and Technology Program).

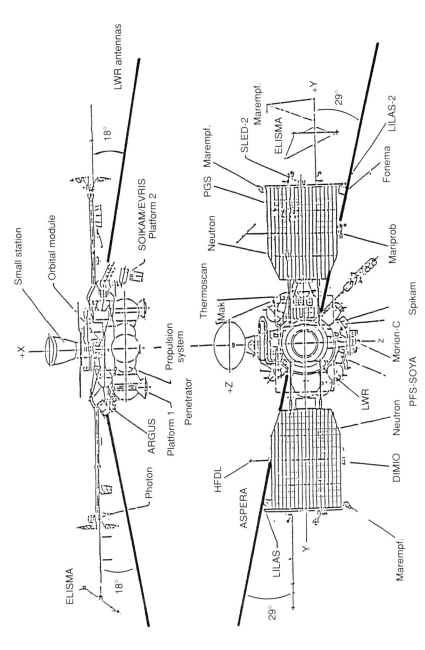

Figure 7.37 MARS 94 Orbiter (courtesy of Max Plank Institut für Aeronomie)

Figure 7.38   MARS 96 project antenna (courtesy of University of Rennes)

the balloon made of a series of 18 metallic cylinders), with the scientific goals of measuring the thickness of the Martian permafrost and subsurface layers and of determining their electromagnetic characteristics, (Laplace *et al.*, 1992). The subsurface depth-penetration capability of this radar must be of the order of 1 km with a resolution of about 10 m. The CNRS (Service d'Aéronomie and CEPHAG at Verriéres le Póuisson and Grenoble) and the Riga Aviation University (Latvia) have built an impulse-radar system including a variable-gain receiver and repetitive-trace coherent integration. Preliminary experiments were carried out on glaciers and on sand dunes to test the equipment and to obtain representative experimental GPR profiles. Then, numerical simulations were made to estimate the performance of the prototype system from specific Martian subsurface models.

Technical solutions

Most of the design of the Mars 96 balloon mission GPR is constrained by the mechanical shape of the guiderope, the allocated mass, the working and storage temperatures, and the limited power available. In these limits, and according to the deep penetration objective, the CNRS and the University of Riga have built an impulse-excitation radar working in the 10 MHz frequency range (Barbin *et al.*, 1995). The guiderope modules house the radar electronics and the power supply (Figure 7.38). A 12 m long flexible dipole consisting of the guiderope as one arm and a thick metallic cable as the other arm forms the antenna. The simple transmitting system consists of short pulse excitation of the dipole resistively loaded in its centre. The antenna, which can be modelled as a leaky open-ended line (Finkelstein *et al.*, 1986), radiates a transient HF field for about 100 ns. Experimental tests show a centre frequency of approximately 12 MHz and a usable bandwidth of 6 MHz, corresponding to a depth resolution of about 12 m in permafrost. At the end of the transmitted pulse, the antenna is connected to the protected receiver which recovers its nominal linear regime after about 250 ns from the beginning of the pulse, corresponding to a blind depth of 20 m in permafrost. The direct 'flash' digital conversion of the complete received signal is made, so that every shot produces a full trace. Thus, it is possible to perform in real time the coherent integration of 256 successive traces to give a 24 dB improvement in signal-to-noise ratio. In the receiver, a time varying gain corrects the geometrical spreading loss, rising from 15 to 60 dB during 3 μs. Practical limitations in the data flow back to Earth drastically limits the amount of data. At the beginning of each night of the ten-days Mars 96 mission, four 40-traces profiles will be performed, separated by some 5 km. The profile length will be on the order of 400 m with a spacing of about 10 m between sounding points.

Experimental results
Numerous profiles have been collected on the Pyla dune (south of the Arcachon basin in France), on the Mont-de-Lans glacier (French Alps) and in Antarctica (Nicollin *et al.*, 1992; Nicollin and Kofman, 1992). The data are processed to remove coherent system ringing and by inverse filtering to improve the resolution and enhance the reflected echoes. On profiles recorded on the Pyla dune (a long dune up to 117 m thick consisting of homogeneous sand overlaying the water table), the echo from the bottom of the dune is clearly visible. Profiles recorded on cold parts (ice below 0°C) of the Mont-de-Lans glacier clearly show the echo from the ice–bedrock interface at depth in the range 50–100 m. Figure

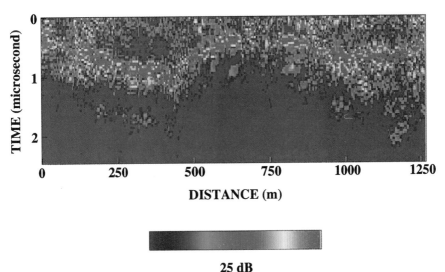

COLD GLACIER

25 dB

Figure 7.39a MARS 96 Project–Radar image of Mont de Lans glacier (courtesy of University of Rennes)

COLD GLACIER

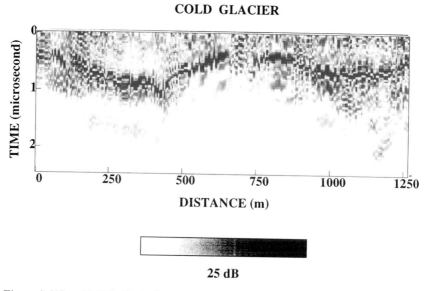

25 dB

Figure 7.39b   MARS 96 Project–Radar image of Mont de Lans glacier (courtesy of University of Rennes)

7.39*a* shows the radar image in colour and Figure 7.39*b* shows the radar image in greyscale. This echo consists of a mixture of reflected (specular reflection) and backscattered (diffuse reflection) waves from an interface which may be considered as a rough surface with respect to the wavelength. Similar results are obtained from profiles recorded in Antarctica, showing the ice–bedrock interface at depth down to 900 m over distances of some 30 km. Profiles recorded on temperate parts (ice at 0°C) of the Mont-de-Lans glacier exhibit widely scattered energy decreasing with depth.

This has been explained by scattering effect in the temperate glacier which can be modelled as a layer of ice containing inclusions of liquid water. Numerical simulations, using a first-order multiple-scattering approximation and the Rayleigh-scattering model, show that, when the reflection at the basal interface is lost in the scattered power, the received power decreases as a function of depth in the same way as the energy decreases in the experimental data. From these results, it is possible to estimate the depth of the ice–bedrock interface for given electromagnetic characteristics of the temperate glacier.

Martian modelling

Geological models of the Martian subsurface show a thick regolith (blanket of debris resulting from impact processes), which could be interbedded with volcanic flows or sedimentary deposits, and which contains a significant quantity of water as ground ice making a permafrost (Fanale, 1976; Carr, 1979). From two models of the planned sites of the Mars 96 mission, several electromagnetic models were built in order to perform numerical simulations. The regolith is represented by a homogeneous porous medium containing a uniform distribution of spherical blocks of basalt. The porous medium is basaltic sands, mixed with Mars atmosphere in the upper dry layer, with ice in the permafrost and with liquid water below the permafrost. Values of the physical parameters are derived from the Viking observations. With a first-order multiple-scattering approximation, we calculated the received power consisting of scattered power in the inhomogeneous medium and power reflected at the different interfaces. These simulations show that, even if the block size is very much smaller than the wavelength (50 times smaller), it is not possible to neglect scattering effects of inclusions in these absorbing media. Nevertheless, if the medium has a tenuous content of heterogeneities (poor blocky regolith) or if it consists of sedimentary deposits, the ground-penetrating radar has the capability to detect the bottom of the permafrost at depths greater than 2 km.

7.9 References

ABRAHAMSON, S., BRUSMARK, B., GAUNARD, G., and STRIFORS, H. C. (1991): Target identification by means of impulse radar *in* SADJADI, F. A. (Ed.): Automatic object recognition. *Proc. SPIE*, (1471), 130–141

ABRAHAMSON, S., BRUSMARK, B., GAUNARD, G., and STRIFORS, H. C. (1992): Extraction of target signature features *in* the combined time-frequency domain by means of impulse radar *in* SADJADI, F. A. (Ed.): Automatic object recognition II. *Proc. SPIE*, (1700), 102–113

ANNAN, A. P., and DAVIS, J. L. (1977): Impulse radar applied to ice thickness measurement and freshwater bathymetry. *Geol. Soc. Canada*, 77.1B

BARBIN, Y., KOFMAN, W., ELKIN, M., FINKELSTEIN, M., GLOTOV, V., and ZOLOTAREV, V. (1991): Mars 96 subsurface radar. Proceedings of ESA international symposium on *Radars and Lidars in Earth and Planetary Sciences*, ESA SP–328, 51–58

BARBIN, Y., NICOLLIN, F., KOFMAN, W., ZOLOTAREV, V., and GLOTOV, V. (1995): Mars 96 GPR program. *J. Appl. Geophys.*, **33**, 27–37

BASTON-PITT, J. D. (1992): The role of impulse radar in the investigation, restoration and change of use of old buildings. Proceedings of 3rd international Masonry Conference, London

BAUM, C. E. (1976): The singularity expansion method *in* FELSEN, L. B. (Ed.): *Transient electromagnetic fields*. Springer, 129–179

BERLIN, G. L., TARABZOUNI, M. A., AL-NASER, A., SHEIKHA, K. M., and LARSON, R. W. (1986): SIR–B subsurface imaging of a sand-buried landscape: Al Labbah Plateau, Saudi Arabia. *IEEE Trans.*, **GE-24**, 595–602

BIGGS, A. W. (1977): Sea ice thickness measurements with short pulse radar systems. URSI Commission F, La Baule, France, 159–163

BISHOP, J. F., CUMMING, A. D. G., FERRARI, R. L., and MILLER, K. J. (1980): Results on impulse radar ice-depth sounding on the Vatnajokull ice-cap, Iceland. Proceedings of conference on the *International Karakorum Project*, Vol. 1, 126–134 (see also 111–125)

BLOM, R. G., CRIPPEN, R. E., and ELACHI, C. (1984): Detection of subsurface features in SEASAT radar images of Means Valley, Mojave Desert, California. *Geology*, **12**, 346–349

BOMAR, L. C., HORNE, W. F., BROWN, D. R., and SMART, J. L. (1988): Determining deterorated areas in Portland cement concrete pavements using radar and video imaging. US Transportation Research Board report TRB NCHRP 304, p. 107

BOTROS, A. Z., OLVER, A. D., CUTHBERT, L. G., and FARMER, G. A. (1984): Microwave detection of hidden objects in walls. *Electron. Lett.*, **20**, 824–825

BREED, C. S., MCCAULEY, J. F., SCHABER, G. G., WALKER, A. S., and BERLIN, G. L. (1982): Dunes on SIR–A images *in* CIMINO, J. B., and ELACHI, C. (Eds.): Shuttle Imaging Radar–A (SIR–A) experiment. NASA Jet Propulsion Laboratory, USA, publication 82-77, 4–52–4–87

BREED, C. S., SCHABER, G. G., MCCAULEY, J. F., GROLIER, M. J., HAYNES, C. V., ELACHI, C., BLOM, R., ISSAWI, B., and MCHUGH, W. P. (1983): Subsurface geology of the Western Desert in Egypt and Sudan revealed by Shuttle Imaging Radar (SIR–A). Proceedings of first *Spaceborne Imaging Radar Symposium*, Jet Propulsion Laboratory, Pasadena, CA, USA, 10–12

BREED, C. S., MCCAULEY, J. F., and DAVIS, P. A. (1987): Sand sheets of the Eastern Sahara and ripple blankets on Mars *in* FROSTICK, L., and REID, I. (Eds.): Desert sediments - ancient and modern. London Geological Society special publication 35, 337–359

BROWN, W. E. (1971): Lunar subsurface exploration with coherent radar. Proceedings of conference on *Lunar Geophysics*, Lunar Science Institute, Houston, TX, USA, 113–127

BUNGEY, J. H., and MILLARD, S. G. (1993): Radar inspection of structures. *Proc. Inst. Civil Eng. (Struct. Build.)*, **99**, 173–186

BUNGEY, J. H., and MILLARD, S. G. (1995): Detecting sub-surface features in concrete by impulse radar. *Non-Destr. Test.*, **37**, (12), 33–51

BUNGEY, J. M., MILLARD, S. G., AUSTIN, B. A., THOMAS, C., and SHAW, M. R. (1996): Permittivity and conductivity of concrete at radar frequencies. International Congress of Concrete in Service of Mankind, Dundee (accepted)

BUNGEY, J. H., MILLARD, S. G., and SHAW, M. R., (1993a): A simulation tank to aid interpretation of radar results on concrete. *Mag. Concr. Res.*, **65**, (164), 187–195

BUNGEY, J. H., MILLARD, S. G., and SHAW, M. R., and THOMAS, C. (1991): Operational aspects of radar investigations. *Br. J. Non–Destr. Test.*, **33**, (12), 599–605

BUNGEY, J. H., SHAW, M. R., and MILLARD, S. G. (1993b): The influence of reinforcing steel on radar surveys of concrete structures *in* FORDE, M. C. (Ed.): *Structural faults and repair*. Technics Press, Edinburgh, Vol. 3, 43–50

CANTOR, T. R., and KNEETER, C. P. (1982): Radar as applied to evaluation of bridge decks. *Transp. Res. Rec.*, **853**, 37–42

CARR, A. G., CUTHBERT, L. G., and LIAU, T. F. (1986): Signal processing techniques for short-range radars applied to the detection of hidden objects. Proceedings of 7th European Conference on *Electrotechnics, EURCON 86*, Paris, France, 641–646

CARR, M. H. (1979): Formation of Martian flood features by release of water from confined aquifers. *J. Geophys. Res.*, **84**, 2995-3007

CARSON, A. M. (1991): Application of radar methods to road assessment. Contribution to Concrete Society working party on NDT: subsurface radar state–of–the–art report

CHAN, L. C., MOFFATT, D. L., and PETERS, L. (1979): A characterisation of subsurface radar targets. *Proc. IEEE*, **67**, 991–1000

CHUFO, R. L. (1992): Noncontacting coal and rock thickness measurements with a vector network analyser. Proceedings of the RF Expo East Conference, Tampa, FL, USA, 41–50

CHUNG, T., and CARTER, C. R. (1991): Pavement thickness and detection of pavement cavities using radar. Ontario Ministry of Transportation report MAT–91–03

CLAASEN, T. A. C. M., and MECKLENBRÄUKER, W. F. G. (1980a): The Wigner distribution – a tool for time-frequency analysis. Part 1. *Philips J. Res.*, **35**, 217–250

CLAASEN, T. A. C. M., and MECKLENBRÄUKER, W. F. G. (1980b): The Wigner distribution – a tool for time–frequency analysis. Part 2. *Philips J. Res.*, **35**, 276–300

CLAASEN, T. A. C. M., and MECKLENBRÄUKER, W. F. G. (1980c): The Wigner distribution - a tool for time-frequency analysis. Part 3. *Philips J. Res.*, **35**, 372–420

CLEMENA, G. G. (1983): Non-destructive inspection of overlaid bridge decks with ground penetrating radar. *Transp. Res. Rec.*, **899**, 21-32

CLEMENA, G. G., and MCGHEE, K. K. (1980): Applicability of radar subsurface profiling in estimating sidewalk undermining. *Transp. Res. Rec.*, **752**, 21–28

CLEMENA, G., SPRINKEL, M., and LONG, R. J. Jun. (1987): Use of ground penetrating radar for detecting voids under a jointed concrete pavement. *Transp. Res. Rec.*, **1109**, 1–10

COOK, J. C. (1960): Proposed monocycle-pulse very–high–frequency radar for airborne ice and snow measurement. *AIEE Comm. Electron.*, **51**, 588–594

COOK, J. C. (1975): Radar transparencies of mine and tunnel rocks. *Geophysics*, **40**, 865–885

DANIELS, D. J. (1980): Short–pulse radar for stratified lossy dielectric layer measurement. *IEE Proc. F*, 1980, **127**, 371–379

DAVIS, P. A., BREED, C. S., MCCAULEY, J. F., and SCHABER, G. G. (1993): Surficial geology of the Safsaf region, south-central Egypt derived from remote sensing and field data. *Remote Sens. Environ.*, **46**, 183–203

DE VEKEY, R. C. (1988): Non-destructive test methods for masonry structures. Proceedings of 8th IBMAC conference, Vol. 3, 1673 (9pp)

DE VEKEY, R. C., BALLARD, G. S., and ADDERSON, B. W. (1989): The effectiveness of radar for the investigation of complex LPS joints. Proceedings of Institution of Structural Engineers/Building Research Establishment conference on *Life of Structures*, Vol. 1, 116 (15pp)

DE VEKEY, R. C., MATTHEWS, R. S., and BALLARD, G. S. (1992): An evaluation of impulse radar for floor and wall assessment in concrete system buildings. Proceedings of CIB '92, Conseil International du Batiment, Vol. 1, 190

DE VEKEY, R. C., MATTHEWS, R. S., and BALLARD, G. S. (1993): Using impulse radar for floor and wall assessments in concrete system buildings. *Construction technology update*. National Research Council of Canada

DELLWIG, L. F. (1969): An evaluation of multifrequency radar imagery of the Pisgah Crater area, California. *Mod. Geol.*, **1**, 889–892

DONG, Z.-B., FERRARI, R. L., FRANCIS, M. R., MUSIL, G. J., OSWALD, G. K. A., and ZHANG, X.-S. (1980): Impulse radar sounding on the Hispar glacier. Proceedings of conference on the *International Karakorum Project*, Vol. 1, 100–110

ECHARD, J. D., SCHEER, J. A., RAUSCH, E. O., LICATA, W. H., MOOR, J. R., and NESTOR, J. A. (1978): Radar detection, discrimination and classification of buried non-metallic mines. Georgia Institute of Technology, engineering experiment station, Atlanta, GA, USA, contract DAAG53–76–C–0112, technical report, Vol. 1

ELACHI, C., ROTH, L. E., and SCHABER, G. G. (1984): Spaceborne radar subsurface imaging in hyperarid regions. *IEEE Trans.*, **GE-22**, (4), 383–388

ELLERBRUCH, D. A., and BOYNE, H. S. (1986): Snow stratiography and water equivalence measured with an active microwave system. *J. Glaciol.*, pp. 225–233

ELLERMEIER, R. D., SIMONETT, D. S., and DELLWIG, L. F. (1967): The use of multi-parameter radar imagery for the discrimination of terrain characteristics. IEEE International Convention Record, Pt.2, 127–135

EVANS, S. (1965): Dielectric properties of ice and snow – a review. *J. Glaciol.*, **5**, 773–792

FANALE, F. P. (1976): Martian volatiles: their degassing history and geochemical fate. *Icarus*, **28**, 179–202

FARR, T. G., ELACHI, C., and CHOWDHURY, K. (1986): Microwave penetration and attenuation in desert soil: a field experiment with the Shuttle Imaging Radar. *IEEE Trans.*, **GE–24**, 590–594

FERNE, B. W. (1991): Non-destructive road survey systems. Proceedings of seminar on *Use of Impulse Radar as Non–Destructive Test of Man–Made Structures*, Loughborough University, UK

FINKELSTEIN, M. I. (1977): Subsurface radar. *Telecomm. Radio Eng. Part 2*, **32**, 18–26 (English translation)

FINKELSTEIN, M., KUTEV, V., and ZOLOTAREV, V. (1986) *Applications of subsurface radar in geology*. Nedra, Moscow, 128

FUJINO, K., WAKAHAMA, G., SUZUKI, M., and MATSUMOTO, T. (1985): Snow stratigraphy measured with an active microwave sensor. Proceedings of IEE international symposium on *Antennas and Propagation*, 671–674

GUNTON, D. J., and SCOTT, H. F. (1987): Radar detection of buried pipes and cables. Institution of Gas Engineers, 53rd Annual meeting, London, UK, Communication 1345

GUO, H., SCHABER, G. G., and BREED, C. S. (1986): Shuttle imaging radar response from sand and subsurface rocks of Alashan Plateau on north-central China.

Proceedings of 7th international symposium on *Remote Sensing for Resources Development and Environmental Management*, Enschede, Netherlands, Balkema, Boston, USA, 137–143

GUSTAFSSON, A. (1992): Object classification by means of pole estimation from impulse radar returns. FOA report C30668–8.4, 3.3, Linköping (in Swedish)

HALL, D. K., (1982): A review of the utility of remote sensing in Alaskan permafrost studies. *IEEE Trans.*, **GE–20**, 390–394

HOLT, F. B., and EALES, J. W. (1987): Non-destructive evaluation of pavements. *Concr. Int.*, **9**, 41–45

ILZUKA, Z., and FREUNDORFER, A. (1984): Step frequency radar. *J. Appl. Phys.*, **56**, 2572–2583

ISSAWI, B., and MCCAULEY, J. F. (1992): The Cenozoic rivers of Egypt: the Nile problem *in The followers of Horus - studies in memory of Michael A. Hoffman*. Oxbow Press, Oxford, 1–18

KADABA, P. K. (1976): Penetration of 0.1 GHz to 1.5 GHz electromagnetic waves into the earth surface for remote sensing applications. Proceedings of IEEE SE Region 3 conference, 48–50

KOVACS, A., and MOREY, R. M. (1983): Detection of cavities under concrete pavement. US Army Corps of Engineers Cold Regions Research & Engineering Laboratory, CRREL report 83–18

KOVACS, A., and MOREY, R. M. (1989): Estimating sea ice thickness using time-of-flight data from impulse radar soundings. US Army Corps of Engineers Cold Regions Research & Engineering Laboratory, CRREL report 89–22

KUNZ, J. T., and EALES, J. W. (1988): Non-destructive evaluation of pavement using infrared thermography and ground penetrating radar. Proceedings of 37th annual Arizona conference on *Roads and Streets*, University of Arizona, Tucson, AR, USA, 29–41

LAPLACE, H., MORELIERE, M., and GROSE, C. (1992): The Mars 96 balloon guiderope: an autonomous system in extreme environmental conditions. Proceedings of international symposium on *Missions, Technologies and Design of Planetary Mobile Vehicles*, CNES, CEPADUES, 417–421

LUKASHENKO, Y. I., MASLOV, A. I., TERLETZKI, N. A., and TSIURUPA, L. A. (1979): Subsurface sounding in the uhf band of lenses of ground water and continental glaciers. *Telecomm. Radar Eng.*, **34**, (4), 94–95 (English translation)

LUNDY, J. R., and MCCULLOUGH, B. F. (1989): Delamination in bonded concrete overlays of continuously reinforced pavement. Proceedings on 4th international conference on *Concrete Pavement Design and Rehabilitation*, Purdue University, Indiana, USA, 221–229

MANNING, D. G., and HOLT, F. B. (1983): Detecting deterioration in asphalt-covered bridge decks. *Transp. Res. Rec.*, **899**, 10–19

MAXWELL, T. A., and PHILLIPS, R. J. (1978): Stratigraphic correlation of the radar-detected subsurface interface in Mare Crisium. *Geophys. Res. Lett.*, **5**, 811–814

MCCAULEY, J. F., BREED, C. S., and SCHABER, G. G. (1986a): Megageomorphology of the radar rivers of the Eastern Sahara. Proceedings of 2nd *Spaceborne Imaging Radar Symposium*, NASA Jet Propulsion Laboratory publication 86–26, 25–35

MCCAULEY, J. F., BREED, C. S., SCHABER, G. G., MCHUGH, W. P., HAYNES, C. V., ISSAWI, B., GROLIER, M. J., and EL–KILANI, A. (1986b): Paleodrainages of the Eastern Sahara: the radar rivers revisited. *IEEE Trans.*, **GE–24**, (4), 624–648

MCCAULEY, J. F., SCHABER, G. G., BREED, C. S., GROLIER, M. J., HAYNES, C. V., ISSAWI, B., ELACHI, C., and BLOM, R. (1982): Subsurface valleys and geoarcheology of the Eastern Sahara revealed by Shuttle Imaging Radar. *Science*, **218**, 1004–1020

MCHUGH, W. P., BREED, C. S., SCHABER, G. G., MCCAULEY, J. F., and SZABO, B. J. (1988a): Acheulian sites along the 'radar rivers' southern Egyptian Sahara. *J. Field Archeol.*, **15**, 361–379

MCHUGH, W. P., MCCAULEY, J. F., BREED, C, S., and SCHABER, G. G. (1988b): Paleorivers and geoarchaeology in the southern Egyptian Sahara. *Geoarcheology*, **3**, (1), 1–40

MCHUGH, W. P., SCHABER, G. G., BREED, C. S., and MCCAULEY, J. F. (1989): Neolithic adaptation and the Holocene functioning of tertiary paleodrainages in southern Egypt and northern Sudan. *Antiquity*, **63**, 320–336

MILLARD, S. G., BUNGEY, J. H., and SHAW, M. R. (1993*b*): The assessment of concrete quality using pulsed radar reflection and transmission techniques *in* BUNGEY, J. H. (Ed.): *NDT in civil engineering*. British Institute of Non-Destructive Testing, Vol. 1, 161–185

MOORE, J. R., ECHARD, J. D., and NEILL, C. G. (1980): Radar detection of voids under concrete highways. Proceedings of IEEE *International Radar Conference*, 131–135

NICOLLIN, F., and KOFMAN, W. (1994): Ground penetrating radar sounding of a temperate glacier; modelling of a multilayered medium. *Geophys. Prospect.* **42**, pp. 715–734

NICOLLIN, F., BARBIN, Y., KOFMAN, W., MATHIEU, D., HAMRAN, S. E., BAUER, P., ACHACHE, J., and BLAMONT, J. (1992): An HF bi-phase shift keying radar; application to ice sounding in Western Alps and Spitzbergen glaciers. *IEEE Trans.*, **GE–30**, 1025–1033

OSWALD, G. K. A. (1988): Geophysical radar design. *IEE Proc. F*, **135**, 371–379

PORCELLO, L. J., JORDAN, R. L., ZELENKA, J. S., ADAMS, G. F., PHILLIPS, R. J., BROWN, W. E., WARD, S. H., and JACKSON, P. L. (1974): The Apollo lunar sounder radar system. *Proc. IEEE*, **62**, 769–783

RADOK, U. (1985): The Antarctic ice. *Sci. Am.*, **253**, 82–89

RANSOME, R. C., and KUNZ, J. T. (1986): Non destructive detection of voids beneath pavement. *Publ. Works*, 52–54

RODEICK, C. A. (1983): Pipeline route selection by ground-penetrating radar. *Pipeline Ind.*, 39–44

RODEICK, C. A. (1984): Roadbase void detection by ground penetrating radar. *Highway Heavy Constr.*, 60–61

SAKAMOTO, Y., and AOKI, Y. (1985): An under snow radar using microwave holography. Proceedings of international symposium on *Antennas and Propagation*, Japan, 659–661

SCHABER, G. G., and BREED, C. S. (1993): Subsurface mapping with imaging radar in deserts of Africa and Arizona *in* THORWEIHE, U., and SCHANDELMEIR, H. (Eds.): Geoscientific research in Northeast Africa. Proceedings of international conference on *Geoscientific Research in Northeast Africa*, A. A. Balkema, Rotterdam, 761–765

SCHABER, G. G., MCCAULEY, J. F., BREED, C. S., and OLHOEFT, G. R. (1986): Shuttle imaging radar: physical controls on signal penetration and subsurface scattering in the Eastern Sahara. *IEEE Trans.*, **GE–24**, (4), 603–623

SCHABER, G. G., OLHOEFT, G. R., MCCAULEY, J. F., BREED, C. S., and DAVIS, P. A. (1990): The 'radar rivers' of the eastern Sahara: signal penetration and surface scattering observed by the Shuttle Imaging Radar *in* LUCIUS, J. E., OLHOEFT, J. R., and DUKE, S. K. (Eds.): Abstracts of the technical meeting of the 3rd international conference on *Ground Penetrating Radar*, Lakewood, CO, USA, US Geological Survey Open-File report 90–414, 61

SHAW, M. R., MILLARD, S. G., HOULDEN, M. A., AUSTIN, B. A., and BUNGEY, J. H. (1993): A large diameter transmission line for the measurement of the relative permittivity of construction materials. *Br. J. Non-Dest. Test.*, **35**, (12), 696–704

STEENSON, B. O. (1951): Radar methods for the exploration of glaciers. PhD thesis, California Institute of Technology, Pasadena, CA, USA

STEINWAY, W. J., ECHARD, J. D., and LUKE, C. M. (1981): Locating voids beneath pavements using pulsed electromagnetic waves. National Co-operative Highway Research Programme, US National Research Council Transportation Research Board, report 237

STRIFORS, H. C., GUANARD, G. C., ABRAHAMSON, S., and BRUSMARK, B. (1993): Scattering of short EM-pulses by simple and complex targets in the combined time-frequency domain using impulse radar. Proceedings of *IEEE National Radar Conference*, Boston, MA, USA, 70–75

SZABO, B. J., MCHUGH, W. P., SCHABER, G. G., and HAYNES, C. V. (1989): Uranium-series dated authigenic carbonates and Acheulian sites in Southern Egypt. *Science*, **243**, 1053–1056

TRANSBARGER, O. (1985): FM radar for inspecting brick and concrete tunnels. *Mater. Eval.*, **43**, (10), 1254–1261

ULABY, F. W., MOORE, R. K., and FUNG, A. K. (1982): Microwave remote sensing -

active and passive *in Radar remote sensing and surface scattering and emission theory*. Addison-Wesley, Vol. 2, 816–921

UNTERBERGER, R. R. (1978a): Radar and sonar probing of salt. Proceeds of fifth international symposium on *Salt*, Hamburg, Northern Ohio Geological Society, USA, 423–437

UNTERBERGER, R. R. (1978b): Subsurface dips by radar probing of permafrost. Proceedings of 3rd international conference on *Permafrost*, 573–579

UNTERBERGER, R. R. (1985): Radar and sonar probing of rocks. *Trans. Soc. Min. Eng. AIME*, **276**, 1864–1874

WAITE, A. H., and SCHMIDT, S. J. (1962): Gross errors in height indications from pulsed radar altimeters operating over thick ice and snow. *Proc. Inst. Radio Eng.*, **50**, 1515–1520

Chapter 8

Equipment

8.1 Introduction

There is an ever increasing range of commercially available equipment for surface-penetrating-radar applications. There are now a number of manufacturers and suppliers of equipment such as GSSI, Sensors & Software, Mala, GDE, Penetradar, Rockradar, ERA Technology, NTT, JRC, EMRAD and others. This Chapter provides a brief introduction to a limited selection of equipment manufactured by various companies.

This selection is not intended as a recommendation or endorsement of any of the products or their manufacturers. Prospective users of equipment must satisfy themselves that their selection meets their perceived needs in terms of application, material environment, speed of operation etc. etc. It is to be hoped that incremental developments in the overall performance of radar systems will result in a more sophisticated and user friendly generation of equipment. The end user requires value for money, lightweight, compact, rugged equipment and easy-to-interpret radar images. The designers and manufacturers should be encouraged to meet these needs.

As a general rule, it is unlikely that one size of antenna will cover a full range of applications and it will be necessary to use large antennas to achieve greater range and small antennas to provide increased resolution. In general, manufacturers of surface-penetrating radar offer a range of antennas which can be used with a single control unit. There are considerable variations in system characteristics and the prospective user should evaluate the manufacturers' stated specifications.

Operators of surface-penetrating radars are required to comply with their country's licensing and EMC requirements and any safety regulations, for example intrinsic safety in hazardous environments. In Europe the equipment should carry the CE mark. The following example will illustrate this consideration.

The power per spectral line radiated by the antenna is of prime importance in judging the performance potential of a particular system. One way of evaluating this is to consider the RF pulse-repetition frequency (PRF) and the effective bandwidth of the RF pulse (applied to the antenna terminals).

For a radar with a 50 kHz PRF and an effective RF pulse bandwidth of 50 kHz–1 GHz, the power to the antenna is spread over 20 000 spectral lines each separated by 50 kHz. Thus a transmitter applying a 100 V peak voltage pulse to a 50 Ω input impedance antenna will apply 200 W peak power and the peak power per spectral line applied to the antenna will be 10 mW.

A radar with a 1 MHz PRF and a 22 V peak-voltage pulse applied to the same antenna will provide the same peak power per spectral line but the peak field strength across the antenna terminals will be reduced by a factor of nearly five, thus enabling the latter to offer more inherent compliance with intrinsic safety requirements.

The overall performance of a radar system can be considered using the following example. Given a comparable radiated power per spectral line, the performance of a radar is primarily defined by the performance of the receiver. Here it is the receiver signal-to-noise ratio and the clutter performance which are of prime importance.

The noise of the receiver can be derived from

$$\mathcal{N}_0 = K(T_a + T_r + L_r T_e)B_n$$

where

K = Boltzman's constant

T_a = antenna noise temperature

T_r = noise temperature of the RF connection between the antenna and receiver

L_r = loss of the RF components (cables etc.)

T_e = receiver noise temperature

B_n = receiver bandwidth

or

$$\mathcal{N}_0 = K(T_a + T_R + L_R T_0(F_n - 1))B_n$$

where

T_o = reference temperature of 290K

F_n = noise figure of the receiver

Most impulse-radar systems use a sequential-sampling diode receiver whose noise figure is generally poor compared with conventional radar systems. Certainly, values for the noise figure of between 10 dB to 30 dB are not unusual.

If a system with the value $F = 30$ dB, $L_r = 6.651$ dB losses, temperature = 290 K and bandwidth = 1 GHz is assumed,

$$\mathcal{N}_0 \simeq 1.38 \times 10^{-23}(290 + 1051 + 4.625 \times 290(10^3 - 1) \times 1 \times 10^9 \text{W}$$
$$\mathcal{N}_0 \simeq 1.85 \times 10^{-8}\text{W}$$

Hence the receiver RMS noise voltage, assuming a 50 Ω input impedance for the receiver, is

$$E_0 \simeq 1 \text{ mV RMS}$$

Depending on the statistics of the noise and the detection criteria, it can be considered that, for a 5:1 signal-to-noise ratio, the signal is given by

$$E_s = 5 \text{ mV RMS or 7 mV peak.}$$

As, in general, most RF sampling diodes are not capable of accepting input voltages greater than 1 V peak in their linear region of operation, the receiver has an effective dynamic range given by

$$D_R = \frac{\hat{E}_{max}}{\hat{E}_S}$$

and, in this example,

$$D'_R = 20 \log\{1/(7 \times 10^{-3})\} = 43 \text{ dB}$$

This value can be increased by reducing the RF system bandwidth or by applying time-varying gain prior to the RF sampling head and thus reducing the peak signal level at close ranges. Typically, a compression of up to 30 dB is practical, thus immediately improving the receiver dynamic range to 76 dB. Note that applying time-varying gain after the RF sampling process does not increase the receiver dynamic range. Alternatively, averaging the RF sampled signal effectively improves the signal-to-noise ratio in proportion to $20 \log \sqrt{N}$ where N is the number of samples. Averaging prior to digitisation can only improve the signal-to-noise ratio to a level defined by the dynamic range of the analogue-to-digital converter which, for a 16-bit module, would be 96 dB. Further averaging of the sampled data would have to take place in software and will slow down the rate of system data acquisition in direct proportion to the number of averages.

For example, a 60 dB improvement in signal-to-noise ratio from a 16-bit analogue-to-digital converter would require 10^6 averages to be performed. A radar with a data-transfer rate of 50 scans/s would be slowed to a rate of approximately 4 scans/day, which is clearly impractical.

Other issues which the user should consider are the radiated pulse fidelity and the rate of decay in time of the latter. It is evident that the radiated pulse should be localised in time and not be spread over many nanoseconds due to multiple reflections within either the antenna or antenna feed system. In practice, a rate of decay in excess of 20 dB per pulse length is necessary for uncluttered radar images.

Additionally, the user should confirm that the receiver does not exhibit nonlinear RF-sampling characteristics and a practical test is, in a clear outdoor area, to use a flat metal plate target of 1m^2 and observe the received signal while increasing the range between the target and antenna. The reflected signal should be localised in time and no spurious echoes should be observed at other times.

It is possible to draw up a 'shopping list' of rule-of-thumb user requirements against which to select a suitable equipment. These are given below but the user is advised to assess performance on a calibrated test rig or site wherever possible.

Table 8.1 provides an indication of suitable pulse durations, centre frequency and achievable depth resolution for a range of target depths given certain assumptions. Any variations on these will change the appropriate selection.

Table 8.1   Suitable equipment parameters

Target depth*	Pulse duration†	Centre frequency‡	Depth resolution§
m	ns	MHz	m
< 0.25	0.5	2000	0.025
< 0.5	1.0	1000	0.05
< 1.0	2.0	500	0.1
< 2.0	4.0	250	0.2
< 4.0	8.0	125	0.4
< 8.0	16.0	63	0.8
< 16.0	32.0	31	1.6

* Depth in a medium loss (< 20 dB/m attenuation) material
† Pulse duration to the halfpower width of the main peak
‡ Assumes a transmitted pulse in the general form of a Rayleigh wavelet
§ Assumes a material of relative permittivity = 9

The following example should enable the user to compare how closely a particular equipment meets a particular requirement. For convenience, all powers are referred to as 1 mW and are expressed in dBm (dB with respect to 1 mW). The procedure is to work through the list checking that the parameters specified are met by the system under consideration. The exercise allows variations to be identified and the user to determine whether such variations can be accepted.

The example chosen is a 1.0 ns-duration radar with a target depth range of < 0.5 m.

The example given should enable the potential user to identify, at least on paper, the salient features of the system to be assessed but in practice a field test

Identifier	System requirement	Specified value	Enter value for system under consideration
A	Pulse duration	1×10^{-9}s	?
B	Peak pulse voltage	50 V	?
C	Pulse interval	1×10^{-6}s	?
D	Peak pulse power	50 W	?
E	Mean pulse power (D × A/C)	=50 mW	= D × A/C
F	Mean pulse power	= 17 dBm	?
G	Receiver bandwidth (−3 dB bandwidth)	1 GHz	?
H	Thermal-noise power = (KTB)	−84 dBm	?
I	Receiver noise power referred to the input (usually a measured value)	−78 dBm	?
J	Receiver minimum detectable signal power (usually 14 dB greater than I)	−64 dBm	?
K	Averaging improvement = 20 log $\sqrt{N}$ where N = number of averages prior to digitisation	10 dB	= 20 log$\sqrt{N}$
L	Receiver minimum detectable signal power	−74 dBm	?
M	Time-varying gain variation prior to RF sampling	30 dB	?
N	Receiver maximum input signal with no time-varying gain (TVG); measured value	+ 13 dBm	?
O	Receiver maximum input signal with TVG = M + N	+43 dBm	= M + N
P1	Equivalent receiver dynamic range = O − L	117 dB	= O − L
P2	Actual receiver dynamic range = N − L	87 dB	= N − L
Q	Antenna transmitter/receiver isolation	−40 dB	> ?
R	Peak receiver signal (D − Q)	+6.9 dBm	= D − Q
S	CHECK IS R ≪ O	YES	?
T	Pulse ringdown per pulse length	> −20 dB	?
U	Spurious reflections (to be less than the dynamic range)	> −100 dB	?
V	Analogue-to-digital converter dynamic range (to be greater than P2) (16-bit)	96 dB	?
W	Data transfer rate to meet P2 requirement (assumes an equivalent pulse interval of 1×10^{-5}s)	0.4 million samples/s	?
W1	Effective sample-acquisition time (pulse interval × number of averages)	1×10^{-5} s	
W2	Data-transfer rate to meet P2, V and W1 (V/W1)	1.6 million samples/s	
X	Data-transfer rate to meet P2 requirement	100 scans/s	?

Identifier	System requirement	Specified value	Enter value for system under consideration
Y	Software:	TBA	?
	processing	TBA	
	file format SEG etc.	TBA	
	archiving	TBA	
Z	Power consumption @		?
	12 V	TBA	
	110 V	TBA	
	240 V	TBA	
AA	Total system components	TBA	?
	Total system weight	TBA	
	Shock/vibration/temperature	TBA	
BB	Maintainability	TBA	?
	Upgradeability	TBA	
	Flexibility	TBA	
	Speed of deployment and use	TBA	
CC	Type of target:	TBA	
	planar: delamination, geological		
	linear: pipe, cable		
	void		
	cylindrical		
	spherical		
DD	Antenna:	TBA	
	parallel dipole shielded		
	parallel dipole unshielded		
	crossed dipole shielded		
	crossed dipole unshielded		
	TEM horn shielded		
	TEM horn unshielded		

on a calibrated site is usually the best method of deciding whether a particular equipment will meet the requirements of the applications.

8.2 Survey methods

An operator of a surface-penetrating-radar equipment must have a fundamental understanding of the principles of the technique as well as expertise in the evaluation of the radar data. Unless this is the case it is more than likely that the equipment will not be used to its maximum capability and both the operator and the client will be disappointed in the outcome. The effect of the material and its structure can significantly change the setup of the equipment and its mode of operation and the correct selection is needed for the site under investigation.

Ideally, an informed and accurate site description should be obtained prior to survey, or failing this an on-site assessment should be carried out. Whenever possible, local geological maps should be used, together, if available with site or structural plans. By these means it is possible to determine in advance the type of equipment, its pulse length or operating-frequency range, recording and signal-processing requirements. Where required, the equipment should be capable of on-site data presentation. There is a significant difference in the equipment needed for general search activities and, for example, road survey where multiple sets of equipment are often employed.

Accurate records of the site are essential to subsequent data analysis and interpretation. The site characteristics recorded should consist of the following whenever the information is available:

8.2.1 Site characteristics

Access – by vehicle/by foot
Grid reference
Ordance survey map reference
Local plans
Aerial photographs
Ground photographs
Construction plans
Historical records
Interior/exterior characteristics
Weather conditions
Time, date

8.2.2 Surface characteristics

Manmade or natural
Site use
Surface flatness
Vegetation
Boundaries
Tracks/paths
Disturbance/anomalies
Surface water/frost/ice/snow

8.2.3 Material characteristics

Manmade e.g. concrete/bitumen/slabs/brick/wood
Natural e.g. grass/soil e.g. clay, clay loam, sandy silt loam etc./rock
Sub surface geology
Bedrock geology

8.2.4 Target characteristics

Description of target
Size of target
Target composition
Depth of target
Date of construction

Various methods can be used to record the site dimensions, ranging from simple tape measurements and careful marking out, to electronic distance measurers of the Total Station type. The latter can be used to log data onto a separate computer. However there is still a need to cross-reference data gathered by the radar and integrate these with an accurate plan of the site. One approach is to use a measurement system which automatically triggers the radar at predetermined spatial intervals.

There are two main types of survey, the linear (B-Scan) or a survey area (C-Scan). The linear survey is in general suitable for road structures where a planar structure is under investigation. Where features such as pipes or cables are sought a B-Scan is often insufficient to establish the line of a pipe, as a single point reflector such as a large stone can cause the same radar image as a scan over a pipe.

A C-Scan is therefore useful as a means of producing a plan view of the area of interest. The general assessment of the area to be surveyed should take into account the dimensions of the target and the grid size should be adequate to resolve its area dimensions. It may be necessary to estimate a maximum spacing which can be reduced over a particular area of interest in order properly to resolve that feature.

The requirements for detection are less stringent than for imaging. For detection a survey grid spacing may only need to be sufficient to cover the area of the target whereas target imaging must meet general sampling requirements and in addition each line must be correctly registered with respect to its neighbour.

Another consideration is the plane of polarisation of the electromagnetic energy. For targets with one large area dimension such as a pipe, the radar cross-scattering section will be larger when the polarisation vector is in line with the pipe. This means that any area that is surveyed with, say, parallel dipoles must be surveyed in orthogonal directions to ensure that no targets are missed. The same principle also relates to crossed-dipole antennas.

The speed of survey must also take into account sampling considerations. A survey wheel is generally used to control either the speed of operation of the radar or to place distance markers on the data record.

The speed of survey is usually governed by the effective data-acquisition rate of the radar. It is appropriate to consider the spatial sampling interval in terms of either the interval between each A-Scan or the number of A-Scans per metre. If the effective data-acquisition rate of the radar system is given by t_a and the spatial sampling interval is given by d the maximum speed of survey is given by

$$V_a = \frac{d}{t_a}$$

At a very small spatial sampling interval, the amount of data that can be generated is large and consideration must be given to the storage medium and whether this has adequate capacity for the proposed survey. Unprocessed B-Scan data are often collected in a 16 bit word format. For example, each A-Scan of 256 samples generates 0.5 kbyte of data. At a spatial sampling interval of 1 cm each metre of ground traversed would create 50 kbytes of data. A total length of 1 km would need a 50 Mbyte file length. For multiple-head radar systems using, say, four radar heads, the data-storage requirements multiply accordingly unless data-compression techniques are used.

A general guide to the spatial interval required is that the spatial sampling interval should be 20% of the linear x, y dimensions of the target.

In general, it is advisable that the radar data be monitored as the survey is carried out.

It is also important that the radar survey can be referenced to a local marking point and that the accuracy of the radar survey is known. The traditional method of achieving this is with a survey wheel which either places markers on the B-Scan or controls the rate of data acquisition.

Unfortunately, such wheels can accumulate errors over a large distance and an additional means of logging known positional information is desirable.

As with most survey methods, some flexibility in investigation procedure is important in order to carry out more detailed or supplementary investigations. Both the client and the survey operator should allow for variations to the planned work.

It is also useful to obtain on site 'ground truth' by means of excavation or endoscopic inspection to validate any calibration information previously made.

The remainder of this Chapter considers specific types of equipment from a variety of sources and was up to date at the time of going to print.

Manufacturers should be consulted as to the availability of the products and their current range. All information supplied in this section was provided by the manufacturer and the reader is advised to evaluate the accuracy of the product and its relevance to a particular requirement.

8.3 Geophysical Survey Systems Inc. (GSSI)

Geophysical Survey Systems Inc. has been the major manufacturer and supplier of radar equipment over the last two decades. Based in Salem, Massachusetts, USA, and owned by OYO of Japan, GSSI offers a range of impulse-radar systems covering a wide frequency range and applications.

The product range is modular with a range of antennas for various frequencies which can be used in conjunction with a range of control units known as the SIR-2 shown in Figure 8.1, SIR-3, SIR-8 and SIR-10. The SIR-10 is described below.

The SIR System-10 is designed to interface with Geophysical Survey Systems Inc.'s entire line of state-of-the-art transducers. It offers ground-penetrating-radar (GPR) technology with full digital control of all setup parameters and multichannel colour display. High-resolution radar profiles are collected by pulling the appropriate GSSI transducer along the line being surveyed. Real-time digital signal enhancement and colour display provide the user with immediate, on-site results. Data can be stored digitally on tape for complete

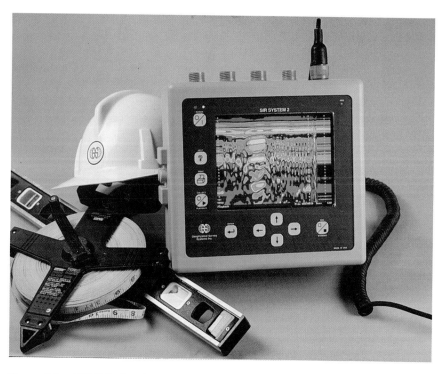

Figure 8.1   SIR-2 Radar (courtesy of GSSI)

post-processing and/or automatically recorded in real time with an optional thermal printer.

The SIR-10 system provides setup flexibility in changing critical data-acquisition parameters, such as gain, sampling, scanning, filtering and transmit rates. With both manual and automatic adjustment of parameters, the SIR System-10 offers a wide variety of control to users at all experience levels. Since all settings can be stored digitally, it offers the user 100% setup repeatability. Digital I/Os allow the SIR System-10 to utilise a wide variety of PCs and PC-compatible devices. While two channels are standard, the SIR System-10 can utilise up to four independent channels, permitting simultaneous operation of transducers with different centre frequencies, signal positions, gain levels, ranges and filter settings. When used with several identical transducers, the SIR System-10 achieves wider coverage per pass, or, a set of varied transducers can be controlled to scan simultaneously for data at different depths. Several transducers can also be configured in arrays to obtain even more detailed information about the subsurface, and can be used in such applications as common-depth-point (CDP) analysis.

The SIR-System-10 includes a digital cartridge tape drive for mass storage of all data, digital settings, alpha-numerics, and time and date information. A full day of surveying data can easily be recorded on a single, small tape cartridge. Taped data can be transferred from the SIR System-10 to an independent PC. SIR System-10 supports the GSSI Model GS-608P digital thermal recorder, enabling the user to create valuable grey-scale data in the field. Analogue linescan recorders are also supported.

Using a high-speed serial link, data can be transferred directly from the SIR System-10's internal tape drive to an independent PC. As the system can also be operated as a standard PC, it can use GSSI's advanced RADAN post-processing software, along with any DOS-based programmes (word processing, etc.). A standard IBM style keyboard can be used in place of the control display unit's internal keypad. External SCSI devices and monitors of any size may also be connected.

A survey wheel input simplifies the process of 'gridding' a site by automatically superimposing distance marks directly on the data, and also eliminates data 'glitches' that occur when transducers are pulled at an uneven rate by recording data only at predetermined distance increments.

With interchangeable 'plug-in' power modules, the user can quickly adapt to any power source requirements. The mainframe can be used in the field with 12 V DC power or with an optional 110/220 V AC power supply. The MF-10 Mainframe is compatible with all GSSI manufactured transducers, including over 15 models that operate from a range of 80 MHz to 2.5 GHz. (Minor, factory-installed modifications are required to operate a 2.5 GHz transducer.) It is also supported by many data enhancement, protection and convenience accessories.

In addition to hardware systems, GSSI offers a software programme suitable for analysing radar data. This program, Radar Data Analyser or RADAN, is a completely menu-driven system which allows the operator to create files, colour

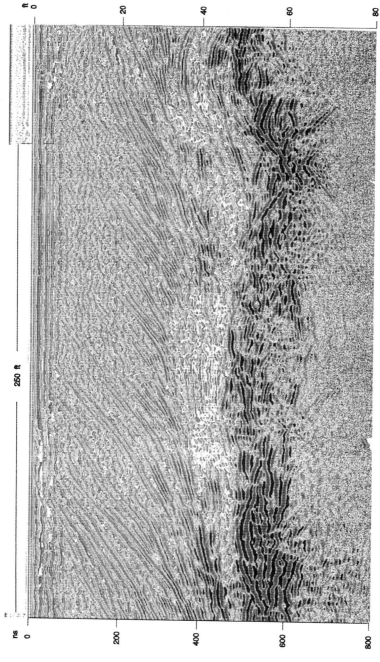

Figure 8.2   Radar images produced using the RADAN program (courtesy of GSSI)

enhance, signal process, modify, and print radar profile data. The RADAN system is especially useful with the interpretation of noisy radar data. RADAN is a completely menu-driven system allowing the operator quickly and easily to choose all of the processing selections and parameters with a mouse, cursor keys or keyboard. Assistance is available with help menus for every function. RADAN includes many data-processing features.

The colour-enhancement module allows user-changeable colour tables and transforms. The user can select from eight preset colour tables composed of 16 out of a possible 512 colours. The colours are selected to represent different signal amplitude values or polarities. The transform thresholds or discrete levels can be described as linear, logarithmic, or exponential. Additional colour and transform tables can be user customised to enhance the data. The radar profiles can be displayed or printed in line scan, wiggle or zoom format. A typical radar image is shown in Figure 8.2.

The SIR System-10 allows the use of the RADAN program to process data collected with two antenna arrays. The operations include adding, averaging, absolute value and RMS functions.

The RADAN program also incorporates a filtering module and includes finite-impulse-response (FIR) or infinite-impulse-response (IIR) filters. Vertical filtering allows scan-by-scan operation in the time domain and incorporates lowpass filters to remove high-frequency (white) noise and highpass filters to remove low-frequency (tilt) noise. Horizontal filtering allows sequential scans to be filtered in individual time-range zones (spatial domain). Lowpass filters average or 'stack' the data to remove white noise and highpass filtering is the same as 'background removal' of horizontal low-frequency noise or ringing.

Deconvolution filtering employs the predictive deconvolution method developed by Peacock and Treital to increase the vertical (time) resolution of events as well as remove multiple reflections. Migration filtering uses hyperbolic summation and Kirchhoff-migration methods that correct the position of an object by collapsing the hyperbolic diffraction pattern into the true shape of the object or slope of the surface. Range-gain filtering employs either automatic gain using an AGC algorithm or user-controlled manual (time-varying) gain with between 2 and 32 breakpoints. Hilbert-transform filtering shows the magnitude (in envelope detection), instantaneous phase, or instantaneous frequency of the data to indicate the energy reflected from an object or how the earth is filtering the radar signal.

Arithmetic functions allow simple operations such as adding or subtracting another file, adding or multiplying by a constant, absolute value square root etc. This can enhance the signal-to-noise ratio or display the difference between processed and unprocessed data.

8.4 Ground Survey Radar (Redifon)

Ground Survey Radar (GSR) is the name given by Redifon to its prototype portable ground-probing radar system. Figure 8.3 shows a picture of the

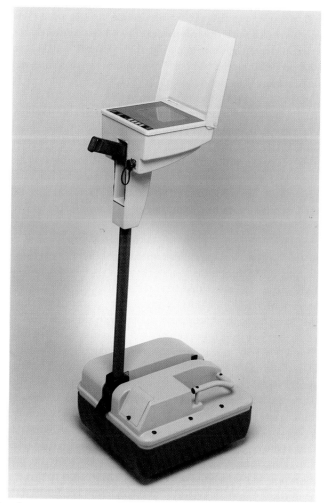

Figure 8.3   Ground-survey radar (courtesy of Redifon)

equipment which is powered by a rechargeable battery contained within a belt worn by the user. The unit was developed in partnership with GSSI of Salem, USA.

The system provides a compact one-man-deployable survey system giving real-time information to the user on a built-in colour LCD display in the usual GSSI time-versus-depth form. Control of the system is achieved via on-screen menus for ease of use. The system thus permits the rapid detection of subsurface features for applications such as pipe and void detection since the radar data are immediately available to the user.

For archiving and data post-analysis requirements the system provides a built-in data storage capability using plug-in RAM cards for robustness. Up to one hour of survey data per card can be captured with the system through the use of data-compression techniques. The data can be decompressed and post processed with GSSI's RADAN signal processing package if required.

It is appreciated within the industry that the interpretation of ground-probing-radar data requires skilled personnel. GSR provides what is believed to be a unique aid to data interpretation in the form of real-time radar signal processing. This is based on a Redifon-designed dual INMOS T800 transputer card running artificial-intelligence (AI) pattern-recognition software written by Redifon. The AI software can be trained on known subsurface features and then used to identify unknown ones during subsequent use. The following section on the AI signal processing provides further detail.

The system is based on GSSI SIR-10 technology. It incorporates a GSSI dual-frequency antenna, radar signal-processing PEC and antenna-interface module. The whole system is controlled by a Megatel single-board PC compatible running software developed by GSSI but modified by Redifon for this application. The data capture and storage is provided on plug-in RAM cards.

The system packaging, user interface and the real-time radar-signal post-processing PEC and software was designed by Redifon.

When using the AI signal processing, the user is presented with a display similar to that shown in Figure 8.4. The reflection pattern is in the standard

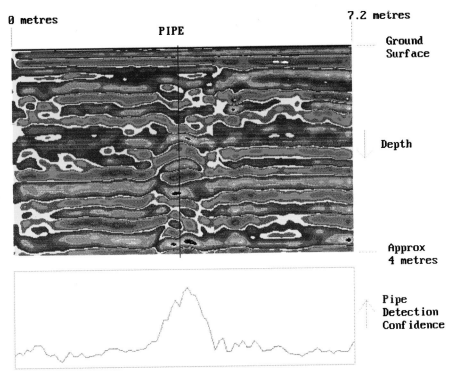

Figure 8.4   Radar image produced by the ground-survey radar (courtesy of Redifon)

GSSI form, with the ground surface at the top of the pattern and depth increasing down the page. From left to right the pattern was created as the equipment was dragged along the ground over a buried pipe. Below the reflection pattern, the output of the AI system (trained to detect pipes) is shown as a measure of confidence that a pipe is present.

The confidence measure is generated in real time by the AI signal processing such that the user sees the confidence level rise as the equipment passes over the object to be detected. Using this the user can, for example, rapidly detect pipes that must be avoided during digging. In an alternative mode the continuous confidence measure is replaced by a simple marker that appears on the screen each time an object is detected.

The GSR can be trained to detect different types of objects by using a menu-driven PC-based training package. The user must first collect reflection patterns from the objects of interest and save these to the RAM card. These are of RADAN format and can be read in and displayed by the training package. The user marks the location of the desired objects and then the package automatically trains the AI to detect these objects. The performance of the trained AI can be assessed within the package before it is transferred back to the GSR via the RAM card.

The RAM card can contain many versions of AI trained to detect different types of objects. Selection of which type of object is detected by the GSR in real-time is then made via the on-screen menus.

The AI signal processing contains a neural network and it is actually this that is trained by the PC package. Neural networks are a form of computing inspired by the human brain which excel at pattern recognition. They are ideally suited to problems such as this where the rules used to arrive at a solution are hard to specify, but example data can be gathered. This is because they are trained on examples rather than being programmed explicitly in sequential steps.

Neural networks are good at generalising such that they recognise patterns with characteristics similar to those of the patterns used for training. To ensure that the neural network concentrates on the most important characteristics, algorithms were developed to extract features from the reflection patterns. These features are ones that are known, from experience, to be important and are fed into the neural network. The combination of the algorithms and neural network provides a system that is more likely to train successfully to give a robust solution.

8.5 Sensors & Software Inc.

Sensors & Software is a Canadian company based in Mississauga near Ontario. It produces two types of radar system; the pulseEKKO 1000 and pulseEKKO IV, as shown in Figure 8.5.

The pulseEKKO 1000 is a lightweight, portable, all-digital ground-

Figure 8.5   Pulse-EKKO IV radar system (courtesy of Sensors & Software)

penetrating radar (GPR) system developed to overcome the constraints of limited performance and portability of existing technology.

It is simple in appearance and has been designed in a modular fashion to give the user maximum flexibility. The system is bistatic (separate antennas for transmitting and receiving). The bistatic design provides the flexibility needed for addressing the wide variety of ground-penetrating-radar applications. The two antennas can be maintained in a fixed arrangement for reflection profiling or moved independently for CMP/WARR velocity soundings or transillumination measurements necessary for tomographic imaging.

A pulseEKKO system consists of six basic components, namely a pair of identical antennas, a transmitter-electronics unit, a receiver-electronics unit, a control console and a personal computer. The pulseEKKO centre operating frequency is selected by mounting the appropriate antenna on the system. All other user-adjustable items are accessed via the personal computer which provides the user interface, graphical data display and data storage.

Resistively damped dipolar antennas are used. These antennas have a ground-plane shield over them which makes them direct energy downward into the material of interest. The antenna radiation patterns are the pattern of a halfwavelength dipole. When the ground conditions are suitable for GPR, the great majority of the radiated signal is transmitted into the ground (typically 90%). The ground-shielded antenna design reduces the impact of nearby metal objects and cables (normally called shielded antennas).

The pulseEKKO antennas are characterised by a nominal centre frequency.

Because optimal radar performance is achieved when the antennas are closely coupled to the ground, the actual centre frequency will vary slightly with ground conditions. Generally, the higher the relative permittivity of the ground, the lower the centre frequency.

Each antenna pair is designed to have a bandwidth-to-centre frequency ratio of 1. In other words, 450 MHz-centre-frequency antennas have useable energy over the frequency range 225–675 MHz.

The pulseEKKO 1000 has a 200 V 800 W transmitter with a 3 dB bandwidth up to 1500 MHz. The transmitter is powered by 12 V and emits a pulse on command (both voltages are provided from the control console).

The power and command from the control console are transmitted to the transmitter unit via a flexible neopreneclad multiconductor cable. The high frequency of the system and unique electronic package allow minimisation of cable interference.

The receiver-electronics module digitises the voltage at the receiver antenna connector to 16-bit resolution. As with the transmitter-electronics module, the receiver is also controlled by, communicates with, and is powered from the control console by an identical (i.e. interchangeable) metallic cable.

The receiver design is such that it acquires the received waveform with very high fidelity. Normally the receiver electronics clip the incoming voltage at a 50 mV level. In other words, any signal larger than 50 mV in amplitude at the receiver antenna is limited at the value of 50 mV. The receiver noise level is nominally around 200 μV for a single stack. If the ambient noise is random, increasing the number of traces averaged to N will decrease the noise level by a factor of $\sqrt{N}$. The present receiver resolution for a single bit after analogue-to-digital conversion is 1.5 μV.

The control console provides the overall management of the transmitter and receiver operation. The control console is a microprocessor-controlled unit which communicates with an external computer over an RS232 communications link. The data-transfer rate on this link is up to 19 200 baud. The master computer passes the system configuration and acquisition parameters to the console which in turn manages all of the hardware functions of the radar system. The console has no user adjustments or knob controls.

For higher-speed data acquisition, an optional fast port is available.

The user interface, the data display and data storage are all carried out using a standard IBM-compatible personal computer operating under MS-DOS 3.3 or greater. The software to control the radar system has been designed to be operable on any level of IBM compatible computer from an XT to a 486. The system operation is set up so that the user interacts with the radar or data display via a menu-driven interface. The menus are self explanatory and, as a result, the system interface is fairly easy to learn.

When the optional high-speed mode of acquisition is used, a high-performance PC with at least 100 Mbyte of disk storage is recommended to make full use of the pulseEKKO 1000 capability.

Data display is in two modes on the screen. One can display the data in individual wiggle trace form which is equivalent to looking at a radar trace on an oscilloscope screen. The alternative is to display the data in section form with a format similar to a seismic-reflection section. When the data are displayed in this mode, the data are scrolled across the screen in either wiggle trace or colour amplitude format. Details of other pulseEKKO software are given below.

Data storage is on any IBM-compatible medium. Generally, the data storage is on a hard disk with backup to floppy disk at the end of data acquisition. It is also straightforward to operate the system on a dual-floppy-disk package with software coming from one disk and data storage on another. Frequently a RAM disk is also used for data storage.

It must be stressed that the data which are stored are the raw data. No adjustments, gains or other changes are made to the data. In this way all data display and subsequent plotting are derived from raw data with no user-controlled parameters between the time of acquisition and the time that data are subsequently displayed. This feature is very important for managing surveys because it removes the user dependence from system operation. Data acquisition is the same no matter who operates the system provided that the antenna configuration is the same.

The EKKO_RUN module provides the user with full control over the pulseEKKO ground-penetrating-radar system. The 'RUN' software comes in standard and optional high-speed configurations. The user configures the hardware parameters such as the time window to be recorded, the sampling interval and the number of stacks to be used. In addition the operator can vary the type of data display on the screen and examine the waveform before actually entering into survey mode of operation. The program is menu driven which makes it easy for even a novice to operate the system. Since the last setup always appears on restart, there is little room for error. Since acquisition details are stored in the data-header file, a history of what was done during acquisition is always available.

The EKKO_PLOT program is the companion to the EKKO_RUN program. In the EKKO_PLOT program the user can recall data stored on disk and display it in a number of forms. At this step, the operator can apply simple filtering, selectively window the data to be plotted and employ a variety of gain functions. The program supports most graphics printers which are compatible with IBM PC MS-DOS operation. High-quality plots are obtained when high-resolution graphics devices are available. For example, all of the data presented in the current pulseEKKO brochures were plotted on an HP Laser Jet printer. The plotting program allows for a wide variety of cosmetic details to be added to the plot including compensation for topography, labelled axes for time and depth or elevation and a labelled position axis. The program allows an individual file to be plotted or a number of files to be merged into a continuous plot. This program, built on many years of experience, provides the user with both screen and hard-copy versions of the same result and is perhaps the most advanced radar plotting program available today.

The EKKO_COLOUR package provides an advanced colour and grey-scale data-presentation capability. The software supports a wide variety of PC colour and black-and-white printers with raster-display capabilities. User-created tables containing up to 256 colours from a palette of 16 million are available.

The colour-plotting program offers the same user features as the EKKO_PLOT program. The rasterised output image is in a GEOSOFT compatible form. Users of the popular GEOSOFT geophysical-data-display software can import radar sections into other display routines.

The EKKO_EDIT program permits the user to edit data files collected with the EKKO_RUN program. In many instances, additional comments or positioning notes have to be added to the data set. In addition there is the occasional operator error or other detail which has to be added or changed in the data files. The EKKO_EDIT program provides a facility to edit existing data files and extract subsets to the data from the original data file.

The EKKO_TOOLS provides the pulseEKKO user with a variety of simple-to-use tools for data manipulation and editing. Processing and display include spatial analysis, FFT filtering and seismic-attribute calculations. Data editing includes, repositioning, commenting, static shifts, and data-file merging. Powerful processes such as section addition and subtraction allow creation of a wide range of data displays with multiple data sets. EKKO_TOOLS run under a user-friendly menu or as stand-alone batch-operable utilities.

The EKKO_SEGY program permits the user to convert standard Sensors & Software Inc. (SSI) format radar files into the seismic-industry standard SEGY data format. The program is simple to use and the operator needs only to enter the data file name and the program automatically creates an SEGY-compatible file. Data in SEGY-compatible format can be imported into any seismic processing program which offers advanced signal processing. An additional capability within this program permits importation of data stored in SEGY format and translation to SSI format.

To facilitate user interaction with the data, the EKKO_SEGY program also provides the capability of data listing to screen, file or printer in ASCII format.

The EKKO_CMP allows the user to apply a constant-velocity stack analysis to a CMP/WARR velocity-sounding file. This program stacks the individual traces in a sounding file with a variety of constant velocities varying from 0.01 m/ns to 0.3 m/ns. The effectiveness of stacking for each velocity can then be displayed with the EKKO_PLOT program. This program allows the user to estimate velocities for various depths in the ground and provides a powerful technique for obtaining depth scales for radar sections.

The EKKO_SYN program allows the user to enter in a layered earth model and compute the synthetic radar response which would be obtained from this model. The program is simple to run and the user only needs to define the layered structure by giving the depth, relative permittivity, and the attenuation in each layer. This information, provided in an ASCII file, is read by the synthetic modelling program. The synthetic modelling program first generates the impulse response of the earth. The impulse response contains all of the multiple reflections which can occur in the sequence and properly accounts for all of the events which can be observed on a radar record. The synthetic program also allows the convolution of the impulse response of the earth with a selected set of wavelets to emulate a real radar-system response. The synthetic program generates files which are compatible with the EKKO_PLOT program which allows the display of the data in a variety of formats. These data files can be also edited and exported using the SEGY program. The EKKO_SYN program provides the user with a powerful modelling capability which aids in the interpretation of ground-penetrating-radar data.

The EKKO_RNG program is a program which allows the user to evaluate the radar-range equation for a given set of model parameters. This program provides powerful capabilities for predicting performance in environments before a survey is carried out. The program follows the full radar range-equation analysis and allows modelling of several target types such as specular reflectors, rough planar scatterers and point scatterers. The velocity or relative permittivity for each medium can be varied along with the attenuation for the host environment. The resulting performance is calculated and provided to the user in several formats.

8.6 ERA Technology

ERA Technology is an independent research and technology company based at Leatherhead, UK. With a strong background in radar and radio-frequency technology, ERA Technology was selected by the UK MoD to develop the first radar system for detecting buried antitank and antipersonnel mines in an active theatre of war, the Falklands, in the early 1980s. In the years following that work, ERA Technology developed special radar equipment for a variety of applications for different clients.

By optimising the radar and antennas, ERA Technology were able to successfully detect various targets for which general-purpose equipment was unsuitable. An example of this approach is the assistance given to the Gloucestershire Police by ERA Technology in detecting the sites of burial of bodies in the basement of 26 Cromwell Street, Gloucester.

The ERA radar system consists of two units: a bistatic antenna with integral radar-head electronics, connected via a multiway cable to a controller unit. A photograph of the ERA radar system is shown in Figure 8.6. This is housed in a purpose-designed rugged case and is based on a 486 DX 66 MHz processor, together with a 540 Mbyte integral hard disk for data storage and built-in control and signal-processing software.

All ERA Technology radar systems operate from 12 V DC and have been designed to be lightweight and field portable. The radars can either be backpack or trolley mounted.

Figure 8.6   SPRscan radar system (courtesy of ERA Technology)

Antennas

A range of antennas is available, in either parallel or crossed configuration, each
with an integral pulse generator optimised for length of the antenna.

All antennas are shielded and are based on the design principle of
continuously loaded elements thus providing an optimised impulse response.
With bistatic antennas, care is needed to achieve low levels of cross coupling as
well as high rates of ring down, thus providing low time sidelobes and an
uncluttered radar image.

As the low-frequency cutoff of all antennas is defined by the length of
the antenna elements and the relative permittivity of the ground, it is
necessary to use different lengths of antenna to radiate different pulse

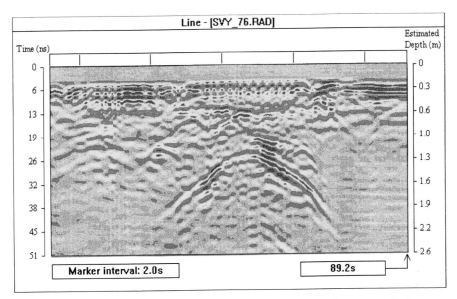

Figure 8.7a   Radar image of storage tank buried under re-inforced concrete slabs (courtesy ERA Technology)

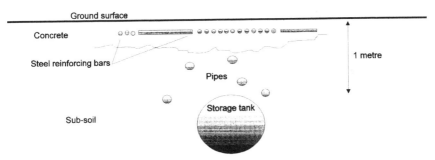

Figure 8.7b   Interpretation of Figure 8.7a (courtesy ERA Technology)

durations. Therefore a range of antenna element lengths of 125 mm, 250 mm, 500 mm and 1 m is used.

Each pair of antennas, i.e. transmit and receive, is housed in a rugged ABS casing which also provides a compartment to house the radar head which can be used with any type of antenna.

Radar head
The radar-head unit consists of the transmitter, receiver, timing controller, analogue-to-digital converter and I/O electronics. The unit operates from 12 V and transmits continuously at a repetition rate of 1 pulse every microsecond. Each transmit antenna has its own built-in pulse generator which is triggered by

First pipe

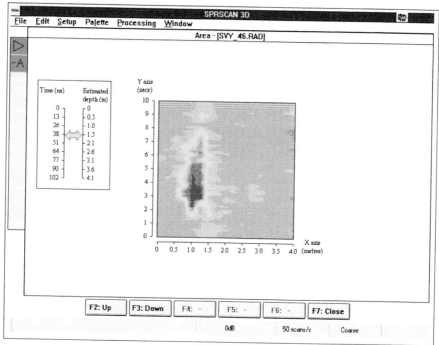

Figure 8.8a   Area scan of buried pipe (No. 1) (courtesy ERA Technology)

the transmitter timing. The high repetition rate enables a mean power to be radiated equivalent to slower, higher peak-power systems, thus removing the need for high voltages in the system.

The basic radar is capable of generating data at a rate of 1 Msample/s. This translates to one A-Scan of 256 samples every 0.256 ms or a scan rate of 3900 scans/s. The receiver RF bandwidth is approximately 2 GHz and the sampling interval can be set between values of 25 ps and 1.6 ns.

A key feature of the ERA receiver is the use of time-dependent gain before the high-speed sampling. The dynamic range of most impulse-radar sampling heads rarely exceeds 60 dB, while the dynamic range of the reflected signals can easily cover 100–200 dB. To overcome the fundamental limitations of the dynamic range of the sampling head and hence improve the sensitivity of the system, the RF subsystem of the ERA transmitter–receiver comprises a low-noise wideband preamplifier and a time-dependent gain module which provides 40 dB of compression in the initial 10–30 ns time period. Thus the equivalent dynamic range of the receiver is nearly 90 dB.

Current-generation 16 bit analogue-to-digital converters can convert data

Second pipe

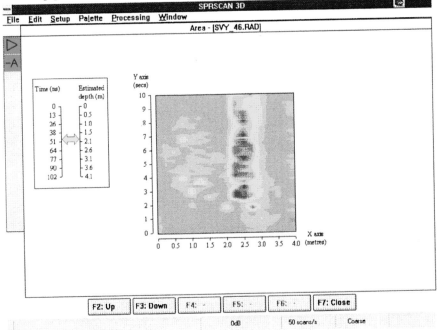

Figure 8.8b   Area scan of buried pipe (No. 2) (courtesy ERA Technology)

every 10 μs; therefore the data rate from the basic radar head is reduced by averaging to achieve compatibility with a 100 ksample/s analogue-to-digital sampling rate. This provides a further 10 dB improvement in effective system sensitivity while offering a maximum operating scan rate of 390 scans/s.

The transmitter–receiver unit is housed in a rugged, field-portable unit approximately 300 mm × 150 mm × 80 mm and contains all the modules to drive the antennas and interface with the controller unit and shaft encoders. The radar head is set up by means of an RS422 interface and the digital output is in fast serial mode.

Radar controller

The radar head interfaces to an IBM compatible personal computer operating under Windows and C. The software to control the radar and the appropriate hardware is based on a 486 DX-66 MHz processor, a 'shock-mounted' 540 Mbyte hard disk, a colour TFT VGA display and a high-speed serial-port data interface which offers a data-transfer rate of up to 100 ksamples/s and scan rate in excess of 120 scans/s. The unit is housed in a rugged field-portable casing which is sealed to IP55 standard.

The controller sets up the radar using an interactive software menu operating

Third pipe

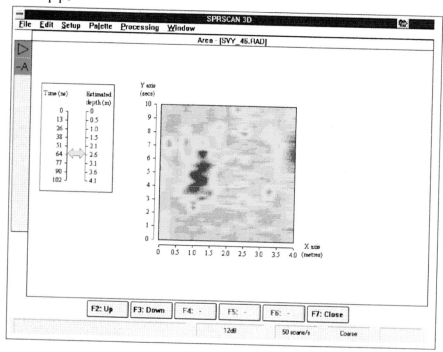

Figure 8.8c   Area scan of buried pipe (No. 3) (courtesy ERA Technology)

in Windows©. In the event that the default values require modification, the user simply selects the required value and the radar head is reconfigured.

The setup module enables the operator to configure the time window, the number of samples and the number of averages. In addition, external control sources such as shaft encoder or external keypad can be called up. The display module allows the operator to check the setup configuration, choose the type of display whether A, B or C Scan, and whether colour or grey scale.

The processing module allows the operator to carry out a number of processing operations on the incoming or stored data. These operations are filtering, multiplying the data by arithmetic constants including range gain, removing background features in the data and detecting salient features in the data.

The data-storage module allows the operator to store data on hard disk. Only raw data are stored and each file is logged with information on the time, date, location and radar-head setup parameters. This procedure ensures that no adjustments or variations are allowed to the initial data, thus removing user dependence from system operation.

The data-output module allows the operator to output data to another

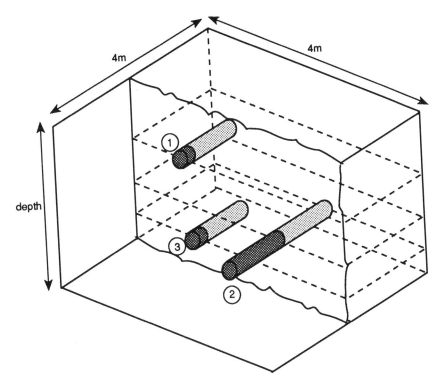

Figure 8.8d Cutaway section of buried pipe test rig (courtesy ERA Technology)

computer for archiving or print copies of the data in either single page or continuous mode.

Examples of data taken using SPR-Scan are shown in Figures 8.7 and 8.8. Figure 8.7 shows a radar B-Scan taken over a buried storage tank through a reinforced concrete road. Figure 8.8 shows a radar C-Scan taken over a buried plastic pipe over an area of 4 m × 4 m. The design concept of the SPRscan is that of a state-of-the-art technical performance combined with user-friendly software to simplify operational set up, thus providing greater productivity in survey operations.

8.7 ESPAR radar system

The ESPAR radar system is manufactured and used by NTT of Japan and consists of a single unit capable of being operated by one man as shown in Figure 8.9.

The ESPAR radar provides a real-time display of buried metallic and nonmetallic pipes at an operating speed of up to 3 km/h using a 2 ns duration

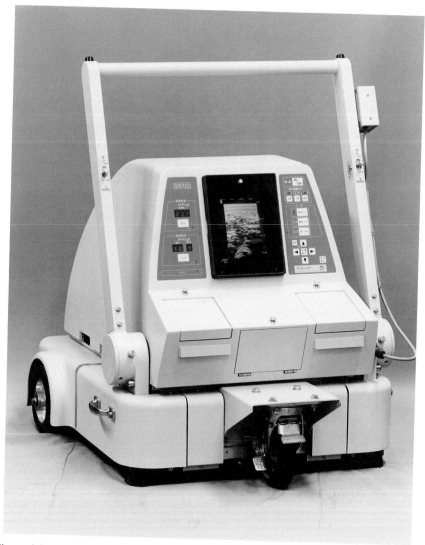

Figure 8.9   ESPAR radar system (courtesy of NTT)

impulse. A maximum operating depth of 1.5 m for a 75 mm-diameter single pipe is claimed and 2 m for a group of pipes. The signal is displayed on an 20 cm colour CRT and the system can measure up to 14 m. The main features of the ESPAR 3 is the ability to calculate the underground propagation velocity of electromagnetic waves from the hyperbolic shape of the cross-sectional image of pulse-radar echoes. The contours of the image are extracted, and the underground propagation velocity is determined from the similarity between the extracted hyperbolic curve and the theoretical curve.

To determine the depth of the buried pipe by means of pulse radar, it is necessary to estimate the underground propagation velocity of electromagnetic waves. A technique based on the synthetic-aperture method requires a large amount of calculation. The ESPAR system uses a new estimation method whose processing time is rapid.

The wave front of the echoes from pipes represents the exact propagation delay time for the distance between the antenna and the pipe when the rising edge of a pulse from the antenna is used as the zero time. The ESPAR radar uses velocity-estimation methods based on the shape of the wavefront known as the 'zero-crossed synthetic aperture'.

The edge-contour information can be obtained by extracting the edges from a cross-sectional image of the radar echoes. The time when the signal amplitude is zero indicates the exact propagation time between the antenna and the target. This is called the zero-crossing time. The amplitude of the observed signal is expressed as 1, 0 or -1, according to the zero-crossing time. The sign indicates the phase change of the original amplitude at the zero crossing.

If the buried-pipe location in the cross-sectional image is (X_0, T_0), propagation time T at each antenna location X can be approximated by the equations

$$T = 2\left(\sqrt{\left\{(X - X_0)^2 + (VT_0/2 + D/2)^2\right\}} D/2\right) \Big/ V$$
$$V = C/\sqrt{\epsilon_s}$$

where D is the pipe diameter. These equations are for a hyperbolic curve. The relative permittivity ϵ_s is the only unknown.

The shape of the theoretical hyperbolic curve with apex location (X_0, T_0) should coincide with one of the extracted edge contours whose apex location is (X_0, T_0). Therefore, if the theoretical calculation uses the real relative permittivity ϵ_s the theoretical hyperbolic curve will have the same shape as the extracted edge contour. Thus the real relative permittivity ϵ_s can be estimated.

Judging the shape similarity, the theoretical hyperbolic curves are calculated at each 1 or -1 co-ordinate regarded as the apex (X_0, T_0). For each calculated curve, the following addition processing is performed. The values of the co-ordinates of the theoretical hyperbola, 1, 1$-$ and 0, are added to the apex (X_0, T_0).

If the above process is repeated for various relative permittivities, the largest sum value is obtained for (X_0, T_0) when the co-ordinate of the theoretical hyperbola coincides with that of the measured hyperbola. Because the addition process is performed at 1 or -1 co-ordinates, the amount of calculation is less than 10% than that of the ordinary method.

A typical radar image is shown in Figure 8.10.

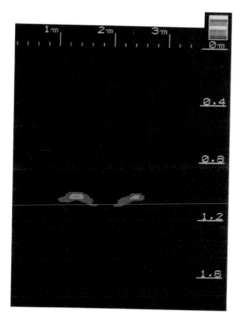

Figure 8.10   Radar image of buried pipes (courtesy of NTT)

8.8 The MALA-RAMAC system

The RAMAC system is a commercially available borehole-radar system which was developed during the early 1980s in the international STRIPA project in Sweden. The system consists of two antenna probes which are used to transmit and receive radar pulses. The electronic equipment to generate and receive the radar pulses is placed in the probes. The measured signals are transmitted to the control unit by optical fibres in digital form. Optical fibres are used to avoid wave propagation along electric cables which would affect the interpretation. Cable winches with cable capacity from 150 to 2000 m are available. The energy for the probes is provided by rechargeable battery packs. The equipment can be seen in Figure 8.11.

The width of the radar pulses is about 20–50 ns. The RAMAC system actually has a timing accuracy of about 200 ps (10^{-12}s). This is essential since the detailed analysis depends critically on the accuracy of timing information during several hours of operation.

A measure is started when the control unit sends a triggering signal to the transmitter, which then emits a short radar pulse. The receiver is activated by a delayed signal and picks a single sample from the measured signal. The whole pulse is then measured by successively changing the delay. The signals may also

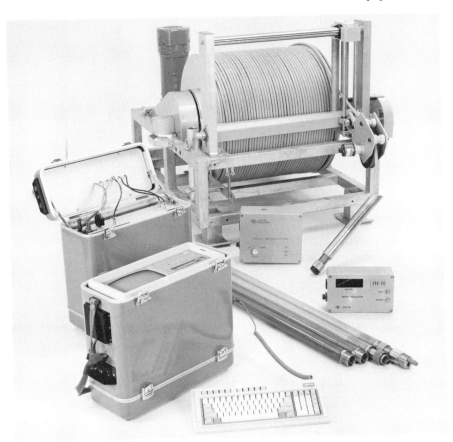

Figure 8.11   Borehole radar system (courtesy of RAMAC)

be stacked a number of times to reduce noise. A typical borehole can be logged at a rate of 120 m/h.

There is little noise in rock when the depth from the ground exceeds 20 m and measurements can therefore usually be performed under extremely stable conditions. This is different from ground-radar measurements and, in spite of the similarities, there are many practical differences between these equipments, concerning both the instrumentation and the analysis.

Data from the RAMAC control unit are collected and displayed on a standard personal computer and stored on a hard disk. Most of the basic analysis can be performed directly in the field, including the tomographic analysis; this is an important feature since poor measurements are discovered and corrected directly. A software package includes tools for signal processing in both time and frequency domains including migration. On-screen interpretation of radargrams by matching of pointlike and planelike objects gives position in borehole and dip angles of structures.

DIRECTIONAL IMAGE IN 90 DEGREES DIFFERENT AZIMUTH

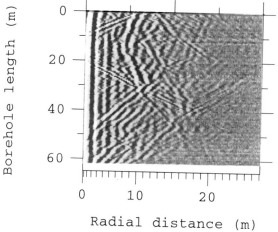

AZIMUTH SHIFTED 90 DEGREES

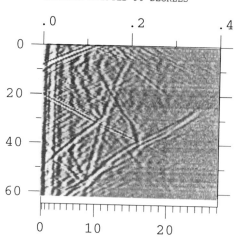

Figure 8.12   *Radar image from borehole system (courtesy of RAMAC)*
a Directional image in 90° different azimuth
b Azimuth shifted 90°

A typical radar record using an antenna centre frequency of 22 MHz is seen in Figure 8.12. The radial detection distance reflects the attenuation in the rock. Prominent fracture zones are seen to intersect the borehole where this distance is drastically reduced. In the image, reflections gathered from the 360° around the borehole are superimposed. Compared with a surface image which represents only the lower halfspace, the borehole image is more complex. By extracting the amplitude of the first arrival along the borehole path for each measurement station this amplitude can be presented as a kind of 'pseudo-resistivity'. Comparing this with, for example, a long normal resistivity log, will show excellent agreement in most cases. The rock volume this energy has penetrated is of course much larger compared with a standard normal-resistivity log. Owing to the greater antenna spacing when using the radar, the curves are much smoother but have the advantage that they can be extracted in PVC-cased holes and in dry holes. Compared with an induction log, the radar also performs excellently in highly resistive areas and can be used, for example, to describe the quality of salt in the mining process.

The ability to perform detailed crosshole measurements is another important feature of the RAMAC system. To describe the geology between two boreholes with good resolution is of vital importance in most site investigations, dam investigations or environment studies. The antennas in the system are independent, which makes it possible to separate them for crosshole studies. Both arrival times and amplitudes are recorded in the data. Velocity tomograms are calculated from arrival times, which depend on the velocity of radar waves, and thus provide a map of the dielectric permittivity in the rock. The dielectric permittivity for water is high compared with that for rock and in many geological environments the bulk relative permittivity will be a measure of the water content. Radar attenuation is normally increased in regions of increased fracturing but high radar attenuation can also be due to other attenuating objects, e.g. conductive ores or clay minerals.

The high timing accuracy of the system makes it possible to perform differential tomography studies. In such, the flow paths in the rock mass can be studied. If, for example, water with small amounts of salt ($>0.5\%$) is injected into discrete fractures, the flow and interaction of fractures can be studied in time by repeated crosshole measurements. Differential studies makes it then possible to calculate the transport time of fluids in the rock mass.

In many cases only a single borehole is available for measurements and the crosshole techniques are then inapplicable. The problem of determining the direction to a reflector in this case has been solved by the development of a directional antenna, which can determine the direction to a reflector within a few degrees.

The centre frequency of the directional-antenna system is 60 MHz corresponding to a wavelength of about 2 m in rock. The main difficulty of constructing antenna is that the borehole diameter is usually much less than the wavelength: the antenna is thus necessarily very inefficient. This may, however, be accepted as long as the dynamic range and the stability of the system are sufficient to measure the signal accurately.

Directional information can be obtained if the induced currents are travelling in opposite directions on the antenna, i.e. essentially in a loop. Such currents will almost cancel which explains the relatively low efficiency of the antenna. This problem is well known in radio techniques. The analysis can be based on transmission-line theory and it is relatively easy to make the antenna sufficiently broadband to leave the pulse undistorted.

Since the RAMAC system uses sampling technique to register a time trace, the timing accuracy of the system is important, and it becomes even more important for the directional antenna since small phase differences are being registered. The time stability allows one to synthesise directional data from four different signals measured with the antenna. One can, in this way, avoid the trouble of having a mechanically rotated antenna in the borehole. The four antenna signals are analysed to extract directional radar signals which are used to synthesise the function of an arbitrarily directed antenna.

By printing radar maps for different directions, one can easily determine the minimum and maximum of a particular reflector. One such example where

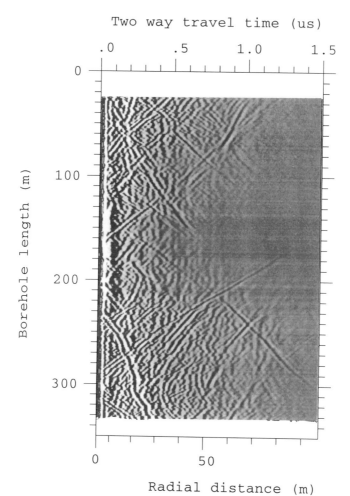

Figure 8.13   Radar image from borehole system (courtesy of RAMAC)

reflectors are seen at maximum and minimum reflection azimuths can be seen in Figure 8.13. Here the curved reflector exiting at '0.2 us' in the lower image is actually a water-filled 76 mm open borehole.

The direction to well defined reflectors can be determined within ±5°, demonstrating that the radar is capable of determining the orientation of a structure from only one single borehole. Note that the structures do not have to intersect the borehole. They may be located remote from the borehole and can still be oriented. On-screen interpretation software is used to analyse the azimuth of reflectors, which can be done in the field.

It is clear that large probing ranges in combination with a resolution of the order of 1 m makes the borehole radar a unique instrument for investigation of structures in rock.

The radar method is most difficult to apply in conductive environments, which present a problem for most electromagnetic methods. The range of the radar will decrease rapidly with increasing conductivity, but measurements with the borehole radar in fairly highly conductive rocks have given surprisingly large penetration depths. It is also interesting that in conductive environments the antenna frequency should be increased to maintain the waveform of the transmitted pulse. This is because the Q-value decreases rapidly with increased conductivity. The Q-value is a measure of how many wavelengths the pulse will propagate in the material before being significantly attenuated. With increased frequency, it will still be possible to keep the Q-value above the critical limit for wave propagation.

The radar method is thus a powerful tool to characterise a rock volume because it makes it possible to see 'through the rock'. Using different measurement configurations, one can achieve a reliable interpretation which is useful in rock characterisation and site investigation projects.

8.9 Technical literature sources

Geophysical Survey Systems Inc., 13 Klein Drive, PO Box 97, North Salem, New Hampshire, 0073-0097, USA

Sensors & Software Inc., 5566 Tomken Road, Mississauga, Ontario, L4W 1PW, Canada

ERA Technology Ltd, Radar Systems Division, Cleeve Road, Leatherhead, Surrey KT22 7SA, UK

Nippon Telegraph & Telephone Corporation, 3-9-11 Midori-cho, Musashino-shi, Tokyo 180, Japan

MALA GeoScience, Skolgatan 11, S-930 70 Mala, Sweden

Other companies also involved are:

EMRAD, Langham Park, Cattleshall Lane, Godalming, Surrey GU7 1NG, UK

Japan Radio Co Ltd, Akasaka Twin Tower (Main), 17–22 Akasaka 2-chome, Minato-ku, Tokyo 107, Japan

Penetradar Corp., PO Box 246, 2221 Niagara Falls Boulevard, Niagara Falls, NY 14304, USA

Cambridge Consultants, Science Park, Milton Road, Cambridge CB4 4DW, UK

Road Radar Ltd, 14535 118 Avenue, Edmonton, Alberta T5L 2M7, Canada

SAIC UK Ltd, Poseidon House, Castle Park, Cambridge CB3 0RD, UK

Index